PESTS

PESTS

HOW HUMANS CREATE
ANIMAL VILLAINS

BETHANY BROOKSHIRE

ecco

An Imprint of HarperCollinsPublishers

HarperCollins books may be purchased for educational, business, or sales promotional use. For information, please email the Special Markets Department at SPsales@harpercollins.com.

Ecco® and HarperCollins® are trademarks of HarperCollins Publishers.

Acknowledgment is made to Annual Reviews, Inc., for permission to reproduce the illustration in the introduction.

FIRST EDITION

Designed by Alison Bloomer

Images © Shutterstock/Florian Augustin, NataliyaF, Aliaksandr Radzko, Svetsol

Library of Congress Cataloging-in-Publication Data has been applied for.

ISBN 978-0-06-309725-4

22 23 24 25 26 FR 10 9 8 7 6 5 4 3 2 1

TO VINCE,
who is my home.

CONTENTS

INTRODUCTION:
A PEST IS _____?

Consider the squirrel.

Many people love squirrels. People cheer for them and smile as their compact, fluffy bodies race over trees and power lines. Every college campus is convinced their squirrels are bolder than any others. I've got a friend who posts a squirrel picture to Twitter every single day, without fail. Squirrels are symbols of fluffy, chipper, charming wildlife come to grace our suburban and urban lives.

Then there's me. Me and F***ing Kevin.

F***ing Kevin is an Eastern gray squirrel (*Sciurus carolinensis*). We call him Kevin for short. He lives in a graceful maple in front of my house. He's a fine, chubby figure of a squirrel. His especially busy tail flicks forward over his back as he trots confidently around my property.

Kevin is my mortal enemy.

This squirrel is the reason I haven't had a tomato from my struggling little garden for at least five years.

I'm a poor gardener at best. But every time the weather warms, optimism kicks in, and I try again. In years past, I would line big pots up on the back porch and plant seedlings with desperate hope.

I've tried most of the usual suspects—basil, zucchini, peppers. But there's a special place in my heart for tomatoes. I have memories from childhood of standing in the middle of my mom's tiny, somewhat dry garden plot in the heat of late July, sneaking cherry tomatoes off the vine and popping them in my mouth. They burst joyfully on my tongue. They were sun warmed and intensely flavorful. The best health food I'd ever tasted.

Every spring, I set out to recapture that perfect experience. I sally forth with hope and plant my tomato seedlings.

Every summer, I am doomed to fail. Because of F***ing Kevin.

There's definitely more than one of him, to be fair. Kevin's probably the godfather of a squirrel mafia. Maybe he's like Batman, with different squirrels donning the mask at different times. To me, though, they're all Kevin.

He owns my yard. He chitters bossily at me from his tree and makes little threatening rushes at me on the sidewalk. But his biggest crimes occur when my precious tomatoes emerge. They swell up, hopeful and green. Every year I look at them and cross my fingers. Just a few more sunny days and those lovely little beauties will be mine. I start planning menus of caprese salad, ratatouille, and salsa.

And every year, Kevin strikes. He selects a nice, plump green tomato and takes a big bite. Suddenly, Kevin recalls that he does not, in fact, like tomatoes. He leaves the perfect green sphere with its tragic tooth marks to rot—making sure to leave it right where I can see it.

Then he does it again. And again. It's an aggressive show of constant optimism. Every evening, it seems, Kevin tries out a new tomato, then remembers that tomatoes suck. He leaves his victims for me to find as a clear sign of his superiority. For five years running, Kevin has taken a bite out of every single tomato in my garden and stuck me with store-bought salsa.

I've tried a lot of things to get rid of Kevin. I put chicken wire around my plants, but squirrel paws (and cherry tomatoes) are smaller than typical chicken wire holes. I sprayed the growing tomatoes with a cayenne-pepper solution to burn his little mouth. He waited, then chowed down after a late summer rainstorm washed it away. I started feeding feral cats on the back porch, thinking a predator or three would keep him at bay. The cats became tame. Two came inside and became our new pets. Kevin added cat food to his diet.

One year I tried the nuclear option. I planted no tomatoes. Instead, I filled pots and cups with jalapeño pepper seeds, hoping Kevin would fall for my devious trick. I fantasized about his squeak of spiced dismay. I pictured a speedy squirrel retreat and the tears running down his furry cheeks (squirrels can't weep, but I can dream).

He never even tasted one. Neither did I. It turns out I'm such a bad gardener I can't even keep a jalapeño pepper alive.

Many of my friends have heard the story of F***ing Kevin. Neighbors know about him; we now call every squirrel in the neighborhood by his name. Most people think it's a silly story of my own incompetence and a squirrel's resourcefulness.

It's also the story of a pest. To other people, Kevin is a simple squirrel, perhaps a little smarter than your average rodent. He's cute, fluffy, possibly even sweet.

To me, Kevin is a constant headache. He makes me feel powerless and foolish. What kind of sciency person, what kind of adult must I be that I can't keep a squirrel away from my tomatoes? Every dead green orb seems to judge me, a silly suburbanite who can't keep a garden alive. And it's all because of F**king Kevin.

I have looked up the average life span of an adult squirrel (six years, give or take two or three), and every year I cross my fingers that this is his last. Maybe this spring he'll come down with mange.

Maybe he'll eat himself to death on my tomatoes. Or maybe I'll break and buy a BB gun.

Kevin is evidence of my inability to control my environment. When we are observing squirrels through the safety of our camera lens, when we have nothing they want, they are adorable wildlife. When they have the temerity to nest in our chimneys, move into our attics without paying rent, or use our gardens as an all-you-can-eat buffet, it's another story.

Squirrel intrusions into our lives are also indicators of animal success. Species of squirrel hang out on every continent except Antarctica (a group of squirrels, by the way, is called a scurry or a dray). From an original diet of nuts and seeds, they've expanded to French fries and bacon. They're one of the few mammals on the planet that can go down a vertical surface headfirst. Most are scatter hoarders—burying caches of food for the lean times in winter. They have highly accurate spatial memories and can pinpoint exactly where they left precious nuts months after hiding them. I have to have someone call my phone at least once a week because I can't remember where I left it.

I admit it. I'm impressed.

Urban charmer or suburban menace? Squirrels are both. Their status doesn't depend on their behavior. They're just trying to live their best squirrel lives. Instead, whether a squirrel is cute or a curse depends on how we see them.

In some places—such as in Scotland—squirrels have gone from persecuted pest to a source of national pride. There, the red squirrel (*Sciurus vulgaris*) is native to the woods and glens. The scientific name might be *vulgaris*, meaning "common," but it doesn't look it. The red squirrel's tufted ears, white tummy, and fluffy tail inspired Beatrix Potter's *The Tale of Squirrel Nutkin* and provided the pattern for thousands of stuffed squirrel toys the world over.

The red squirrel's history in Scotland has been filled with ups and downs, says Matthew Holmes. A historian of science at the

University of Cambridge in England, Holmes has tracked the history of human opinion on animals such as squirrels and sparrows. At first, he explains, forests weren't primarily seen as habitat. People didn't view a forest as a place where things lived, unless they wanted to hunt those things. Instead, forests were trees, and trees were wood to be used in shipbuilding, fires, and more. But as people cut down the local forests, the things in them ran out of places to call home. The red squirrel, which spends most of its time in the trees, began to disappear. By the end of the eighteenth century, no fluffy red tails were to be seen in most of Scotland.

The squirrel was saved in Scotland not by conservationists, but by aristocratic fashion. Around 1778, Elizabeth Scott, Duchess of Buccleuch, saw some red squirrels on her English lands and decided that she had to have them back in Scotland. She wasn't alone. Other aristocrats felt the same—the squirrels went well with the artfully disheveled deer parks they were building on their estates. The lords and ladies weren't necessarily inspired by a charitable sense of conservation—that idea didn't really exist yet. The aristocrats just liked the look of the landscaping—and the money it could bring in from hunting tourism in their new parks. As rich people were more and more able to separate themselves from the worst the wild could send, it became much easier to appreciate its beauty. They brought back squirrels and their habitat in their new enthusiasm for nature.

The animals took advantage, and the squirrel orgy began. They damaged local forests. They gobbled up bird eggs. This was a crime of the highest order to the Victorians, Holmes says, because if they took bird eggs, there were fewer left for the avid Victorian egg collector. Within a century, the red squirrel seemed doomed to a life of dodging gun-toting foresters and angry naturalists.

Until the mid-twentieth century, red squirrels remained unpopular bird-egg munching villains. But as the century wore on, the Scots (and the English) saw their native red squirrel come under attack from the Americans. Not people. Squirrels. *Sciurus carolinensis*, our Eastern gray squirrel, was introduced from the Americas to English and Scottish woodlands in 1876. An army of Kevins quickly took over with a truly American sense of Manifest Destiny. The squirrel also carried squirrelpox, which gray squirrels can resist but reds cannot. Fluffy red tails began, once again, to disappear.

The former red squirrel pest became a cause for conservation concern. "We used to hate the red squirrels, but now [that] they're outcompeted . . . we side with them," Holmes says. "All kinds of weird nationalism and xenophobia comes in." Sure, the red squirrels were bad. But American squirrels were worse. They weren't Scottish, or even from the United Kingdom. They probably chittered in bad accents.

Now, the red squirrel is an icon in the British Isles. It graces teacups and mugs. There are breeding programs and protected areas. There are societies devoted to the squirrel's preservation. The red squirrel has joined the ranks of persecuted wild things.

It's beloved.

FROM UNWANTED TO protected national symbol, pest is all about perspective. Much like buttercups and dandelions. They are weeds when bursting out of your expensive seeded lawn. But when gathered into a bouquet and handed to you by a grubby pair of five-year-old hands, they are the most precious flowers in existence.

"Pest" may seem like a bit of an offensive label for something as cute as a squirrel. It puts them in the same category as rats, mice, or pigeons. All of these are animals that aren't staying in what we've decided is their place. A squirrel in a tree is adorable. A squirrel in your garden, or nesting in your roof, is an annoyance. Something that we should, at the least, control, and at the worst, eradicate.

This means that the idea of a pest is very much in the eye of the beholder. "I remember as an undergrad taking a course in sustainable agriculture," says Philip Nyhus, a professor of environmental studies at Colby College in Maine. "The professor pointed out the terminology of pest and [asked], 'Why is this plant a pest and this other one that looks just like it is not a pest?' Because somebody has decided that one plant is useful to eat . . . or has some economic value. And the other one's essentially impinging on that." The impression the professor made was lasting. Nyhus is still fascinated by human conflict with animals.

Some conservationists might say a pest disrupts a natural ecosystem. Not everyone would agree. English ivy, after all, lends

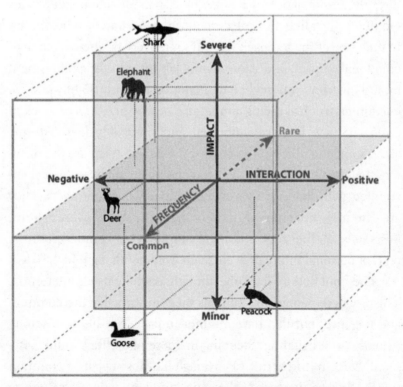

This figure, from Nyhus's 2016 review article "Human-Wildlife Conflict and Coexistence," shows the different ways that people react to the animals around them.

gravitas to a facade, but strangles native plants and rips bricks apart. An outdoor house cat is snuggly little Fluffy in the sheets and a murderer in the streets. Cats kill between one and four billion birds a year in the contiguous United States. But most cat owners get pretty upset if you start calling their beloved Fluffy a pest. The distinction between "pest" and "not pest" is entirely subjective.

Nyhus has been able to pin it down a little. He has categorized human-wildlife conflict into eight quadrants on a graph. On the horizontal axis is how positive or negative an encounter with an animal (for the purposes of this book) or plant is. On the vertical axis is how intense that interaction could be. On the third axis—extending forward out of the page—is how frequent an encounter with a species is. He can then put animal encounters into the eight three-dimensional potential cubes of human judgment. For me, for example, snuggling a wombat on a trip to Australia would end up in the cube at the top back right. It's rare, very positive, has a high emotional impact, and is on my bucket list. On the bottom right toward the front, you might have common but only slightly positive encounters, such as seeing a robin in a nearby tree (or a peacock in a zoo). It's nice, but it's not unusual, and it probably doesn't deeply affect someone's life. In the cube at the back top left are the meetings that are rare, negative, and can end badly, such as physical encounters with sharks (very, very rare), grizzly bears, or tigers. They are extremely uncommon (most of the time, sharks swim right by, bears go the other way, and there's no direct encounter), but the odds of personal injury or even death from conflict are high.

The front bottom-left cube, though, is where it gets interesting. This is where people and animals meet in ways that are common and negative, but that have a minor impact. Not life threatening (usually), but irritating. Stepping in goose poop. Rats in the basement. Mice in the barn. "On average it's those negative common species that would include the agricultural animals and would include those things in your garden," Nyhus says.

The animals on the left side of Nyhus's graph are in for a rough time. We often respond to negative interactions with nature by killing the offender. People call the interactions we have with these animals "human-wildlife conflict," and we are happy to start the fight. We'll fence them out, fence them in, relocate them, and if all else fails, bring out the gun and the poison.

Sometimes it's not just out of irritation. Some of the objects of our wars are apex predators, like the lions, tigers, and wolves that haunt the back upper left of the graph and populate our cautionary fairy tales.

Pests are a step or two below these wilderness Disney villains. Most of the time, pests don't consider humans a potential snack. If a pest does cause death or injury, it's often by chance. If a deer and a car collide, a coyote eats a cat, or mouse poop spreads disease, the animals were not out to harm people.

These generally irritating critters don't tend to cause us the intense fear and anxiety that lions, grizzly bears, or sharks do. They don't directly harm us most of the time. They harm our stuff— snacking on our pets, nesting in our car engines, and pooping on our statuary.

We view their depredations with annoyance. Humans have spent so much of our history trying to keep the weather and the world out of hearth and home that when nature outeats, out-breeds, and outsmarts our best efforts, it's no wonder we end up so angry.

Pests are a problem as old as ownership.

When people were fully immersed in the natural world, had no houses, and grew no crops, the word "pest" probably didn't have a lot of meaning. Other animals were competition, or threats. Birds pecked at the sweet berries we craved. Predators with bigger teeth might have driven us away from that tasty carcass we were about to scavenge. We were part of the natural world, and that included the rivalry for resources.

For a pest to be a pest, people have to have a sense of ownership over stuff they want to protect, and they need to see the difference between what is theirs and what isn't.

Luckily for the pests of the world, people have been hoarding stuff for a very, very long time. We've had storehouses full of grain, haunches of meat, and baskets of fruit. We've left behind food scraps, poop, animal fur, and hair other animals could eat or use. Sure, one creature's trash is another's treasure. But animals helping themselves to our carefully hoarded food? You mess with our food, you mess with us.

People didn't just store things, they also began to settle down. By building shelters, humans began to alter the environment around us. Along with our first huts, we built a humancentric ecosystem. A system other animals wanted into, and that we wanted to keep them out of.

When people raised in Western education systems think of ecosystems, they might remember a school science class. The teacher shows a forest and then a food web. The soil supports the trees. The trees convert air and sunlight into sugars for energy and growth and provide places for songbirds and squirrels to live. The squirrels and birds feast on the plants, and the foxes and owls snack on the squirrels and songbirds. Bugs and fungus decompose the leftovers. It's the circle of life.

Ecosystems, though, are more than food webs. They're the sum total of every organism in a place, and how those organisms interact with their habitat and with each other. They're not just deserts or forests or tundra. A city, town, or farm is also an ecosystem. Ecosystems can be environments that humans create, often without any idea we are doing it.

Every ecosystem has niches—places where organisms can thrive. Some are prime real estate—tropical forests, coral reefs, or fertile marshlands. Others are for more rugged critters. Hot, gaseous hydrothermal vents bubbling up from the sea floor might not look like a happy home to us, but deep-sea crabs and tube worms would beg to differ. A sewer might be too much for a discriminating salamander. But a rat is not so picky.

Humans aren't the only animals that create ecosystems other creatures can use. Remoras live in the small swimming ecosystem of a shark, nibbling its cast-off skin. Beavers build dams (which some humans have cause to curse at) and create ponds that benefit species from waterfowl to moose. Caves full of bats produce mountains of guano—a rich menu for microorganisms and fungus. Those decomposers feed small bugs, which feed cave crickets. An entire ecosystem is based around a bat's behind.

Humans—like beavers or bats—aren't islands unto ourselves. We alter the environments around us. We build ecosystems of concrete and brick and steel. We dig tunnels in the earth, plant succulent, tasty vegetables in gardens, and create dark, shady nooks under roofs. We even create islands of safety, killing off predators that might eat our pests.

Human-built ecosystems—from urban concrete jungles to giant monoculture fields of canola—are invitations to animals in need. A glamorous coral-reef lifestyle it is not. But an organism can certainly make a good living, if they have the right skills. So species moved into our angular, rocklike ecosystems. They started making their living off the things we ignored, left out, or left behind.

EARLY SIGNS OF vermin control are everywhere, says Rowan Flad, an archaeologist at Harvard University. When looking for signs of prehistoric pest control, don't look for things that end up in museums, he says. Instead, look for latrines and garbage heaps. "The reason you put a trash pit outside your house, and not inside your house, is because you want the pests out there," he says. "And you want the decomposing crap to be over there, not inside." Every time scientists find an early crapper or trash can, they find evidence that humans didn't want to deal with the flies, rats, and other animals associated with their waste.

Walls and fence posts could also serve this function. When archaeologists find these defenses, Flad says, they could be against other humans—but they could also be against animals trying to get at human food. A fence can keep both deer and other humans from your garden, and a wall is proof against bandits and pests alike.

This demarcation between people and the natural world they lived in meant welcoming some animals in—the animals that became our dogs and pigs and sheep and cows. It also meant that other animals—predators and pests—began to be seen as unwelcome intrusions into our new and improved lifestyle. And those intruders had to go.

Archaeologists have found ceramic mousetraps that are more than four thousand years old in the cities of Bampur, Mohenjo-Daro, and Mundigak, ancient cities in what are now Iran, Pakistan, and Afghanistan, respectively. The residents of the ancient cities molded clay mousetraps with strings that controlled a sliding door. They looked a bit like tiny mailboxes (with air holes).

Stories of pests have been passed down from antiquity. Anyone who has read the Old Testament book of Exodus will remember the ten plagues that God sent to Egypt. The rivers turning to blood and the dreaded killing of firstborn children were the most dramatic, but other bizarre weather events included rains of flies, gnats, and frogs. In the book of Samuel, when the Philistines take the Ark of the Covenant, they are afflicted with a plague of mice and "tumors" (whether these were bubonic plague lumps or hemorrhoids is up for debate). When they returned the Ark, they had to send with it five golden "images of your tumors and images of your mice that ravage the land." An extra payment to go with the pain. In Aesop's fables, stories allegedly told around the sixth century

BCE, a farmer sets out a net to catch the cranes dining on his grain. Instead, the net catches a hapless stork. The bird pleads for his life, insisting he's not a crane, he just hung around with the wrong birds. The farmer has no pity. If the stork associated with those nasty cranes, well, he deserved the same fate.

As time went on, the divide between animals we loved and animals we loathed sharpened, especially in Western culture. A house or town wasn't just about shelter from the cold and wet, says Harriet Ritvo, a historian of science at the Massachusetts Institute of Technology. It was about people "defending themselves from all the things that nature could do to them." Nature was out there, cold, wet, very muddy, and full of sharp rocks. It wasn't a place you escaped to, it was somewhere you tried to escape from. Ritvo has studied how people have viewed the environment around them over time. "It's an artifact of [Western, often white] modernity that people feel really able to appreciate the natural world as a source of beauty and inspiration and something you go in pursuit of, instead of something that you hide from."

Up until about the Industrial Revolution in Europe, she explains, pests were a fact of life. Most barns had rats, houses had mice, fields had crows, and henhouses had foxes as a matter of course. People devoted a lot of time and money to keeping them out. But they were used to feeling vulnerable, used to feeling that true control could never be had. After all, no one could keep out every pest.

But as human technology improved, Ritvo explains, "the obstacles to all kinds of things presented by the natural world diminished." Bogs were drained, wild forests were cleared. The tops of the tallest mountains felt the imprint of our intrepid boots. In the process, we became better and better at keeping nature, in all its hairy and feathered glory, out. Our sense of control increased. Humans—white European humans in particular—were masters of all we surveyed.

Unless, of course, we weren't.

The concept of "pest" has a lot to do with our sense of our own power, Ritvo explains. As our sense of control increases, an animal may go from nasty pest to revered icon of nature. Like the red squirrel. Or the wolf.

WOLVES (*CANIS LUPUS*) are predators. They get top billing as the Little Red Riding Hood–munching hairy nightmares of our childhoods. They have certainly killed their fair share of humans, but, far more often, the gleaming white teeth of storybook fame aren't coming for us. Instead, they are coming for the sheep we guard or the deer we want to hunt for our supper. In many areas, wolves still go after sheep and cattle. Why hunt a fast-running deer, when the slow, stupid, fluffy sheep aren't going anywhere?

Throughout history, wolves were a source of fear and loathing. Killing wolves was a way of imposing order over chaos, civilization over barbarism. According to William of Malmesbury (writing in the twelfth century), after Edgar took the throne of England in 959, he could have continued demanding from the Welsh a payment of gold and silver imposed by his uncle, Aethelstan. Instead, he demanded three hundred wolfskins a year.

Wolfskins were probably not having a fashion moment in Edgar's court. The new tribute had a purpose—persecution of a pest. More wolfskins meant fewer wolves, and fewer wolves probably meant more sheep grazing the grassy hills of Wales. The wolfskin tribute stood for only three years. After that, William says, Wales seems to have run out of wolves.

The English nobility continued to hunt wolves across the rest of the island for centuries. The slaughter worked. By the mid-sixteenth century, Little Red Riding Hood could have traipsed through England and Wales in safety (though she might have wanted to avoid Scotland and Ireland for another two centuries).

Across the Atlantic, in the United States, starving Jamestown colonists faced wolves in 1609, and a Massachusetts Bay colonist beat one off with a stick in 1621. Colonists expanding across the continent shot them to protect their cattle and sheep. But they also killed them because wolves ate deer, elk, and other animals that humans wanted for their own meals.

This arises out of a utilitarian view of both animals and landscape, says Adrian Treves, head of the Carnivore Coexistence Lab at the University of Wisconsin–Madison. In a utilitarian view, people are not part of nature. Nature exists, complete with animals, to serve people. In that view, Treves says, anything that eats livestock—wolves, coyotes, anything—is a "nuisance" or a "problem" animal. A problem that needs to be solved by getting rid of the species in question.

That sense of dominion, the idea that the world is ours for the taking, runs through a lot of Western history. "I would say the whole human nature dichotomy has been argued to be a Western construct," says bethany ojalehto mays, a cultural psychologist at Cornell University. "So much of our life is constructed in a way that what is human is partitioned from what is natural, and we have created a lot of things that make sense of that divide." Aristotle put humans above the animals, Descartes declared them no more than living machines.

Colonists had this utilitarian view. But not everyone does. Some Indigenous groups see wolves (*Ma'iingan* in the Anishinaabe languages) as kin, fellow predators in the world. In the Anishinaabe creation story, when the Creator made the first human, the man was lonely. So the Creator made not a woman but a ma'iingan. Man and ma'iingan lived together, traveled together, and together they named the world. Finally, the Creator said that the two would have to go their separate ways. But the Creator also told them they were tied together for eternity, that "whatever happened to one would happen to the other. Each would be feared, respected, and misunderstood by the people who would later join them on earth."

To the colonizing people, wolves were, and sometimes still are, considered both predator and pest. They cut into ranching profits with their carnivorous ways. Into the twentieth century, states and the federal government gave bounties to hunters to trap them, poison them, and hunt them down with dogs. (States offered $20 to $50 per wolf, about $280 apiece today.)

Bounties against wolves continued until the wolf gained protection with the Endangered Species Act of 1973. By then, wolves had been almost eradicated from the lower forty-eight states. Humans had won, defeating the predator and the pest. Little Red Riding Hood could have embarked on a world tour.

And then, something strange happened. The colonists became confident victors. Wolves became vulnerable. The balance of power had shifted. Wolves, Ritvo explains, "[stop] being a pest when they can't hurt you anymore." Settlers had killed off so many that our livestock were no longer in any danger, and the deer bred in peace.

With our growing sense of confidence, Ritvo says, our perspective began to change. As other animal populations exploded, scientists began to realize the importance of wolves, grizzly bears, and other apex predators in the environment. People never saw wolves and forgot the fear and frustration of previous generations. Wolves became pure, admirable, and noble. Like the red squirrel, they went from pest to unjustly persecuted, helpless victims of human greed.

As the twenty-first century arrived, Americans became powerful enough to be generous. Wolves have been reintroduced to areas such as Yellowstone National Park, and are striding into places like Colorado, Idaho, and Montana.

"Most [people] have a pretty positive perception of wolves because we don't have to deal with them," Nyhus says. Ranchers, however, aren't as enthusiastic. Wolves still take out cattle and sheep. Everyone may benefit from the beauty of wolves—but it's only those who live with them who pay the price.

Right now, that cost is managed by relatively low numbers of wolves—and a compensation program for ranchers whose livestock go missing. But to some, any wolves are too many. When the U.S. Fish and Wildlife Service removed wolves from the list of Endangered and Threatened Wildlife on January 4, 2021, hunters jumped at the chance. By the end of February, a state-planned wolf hunt in Wisconsin to kill 119 animals blew past the target, killing 218 wolves in a mere sixty hours, over the strenuous objections of the local Anishinaabe peoples, whose ancestral lands are located around the Great Lakes and who include the Ojibwe, Potawatomi, Algonquin, and other tribes. Probable and confirmed wolf depredations in Wisconsin in 2020 totaled 147 animals.

Wolves are now the subject of positive press, tattoos, and really unfortunate T-shirt designs. They are symbols of the natural world. But only to some, and only as long as they don't get too numerous. They exist on our sufferance. Wolves are beautiful as long as they have no power to kill our livestock. As long as they know their place.

AS WESTERN SOCIETIES wall themselves off from the natural world, its denizens fall into two camps: the ones we rarely see and the ones we see too much. The rare animals get appreciation. They are beautiful, natural, and usually far away. The common ones, on the other hand, are so common that in the best case our eyes pass right over them. In the worst case, they intrude on our consciousness and our lives. They become pests.

The animals that we live with, that we can't control, earn increasing amounts of our wrath and disgust. They are now often associated with poverty and poor infrastructure. People view pests with the same disdain they have for unkempt yards and dirty homes.

But they are also, in a way, signs of animal success. "Humans are very good at taking up space, we're a very successful species overall," says Dawn Biehler, who studies the intersection of animals and social justice at the University of Maryland, Baltimore County. "These animals piggyback on our successes to thrive in a lot of different kinds of spaces, and part of what they're especially good at piggybacking on is social inequalities."

Biehler is the author of *Pests in the City: Flies, Bedbugs, Cockroaches, and Rats*, which delves into the connection between urban wildlife and human inequality. She doesn't blame a pest for being itself. "It's not their fault, the way they are. They are just fulfilling their niche requirements, the things they need to do to survive," she says.

Although our infrastructure is designed to keep nature out, every facade has its cracks. Into our carefully hidden sewers, our golden corn mazes, and our dusty attics, crept the pests. We had become successful at keeping nature out. If it got in anyway— didn't that mean that we had failed?

Areas of poverty are areas where those cracks in our infrastructure become visible. Poor neighborhoods, in the United States and well beyond, are often the last served by trash collection and sewer maintenance. They have older buildings, filled with holes and occupied by people too poor to maintain them. Sometimes public housing projects are built cheaply and not maintained, or housing is controlled by rapacious landlords who have little interest in the welfare of their tenants.

Trash piles up. Sewage spreads. Fences collapse. Holes and hiding places abound. Pests move in. "It's really interesting to me how [pests] take advantage of the ways that different groups of humans have treated each other very poorly," Biehler says.

When pests move into these habitats full of delicious opportunities, it's the people who get the blame. The animals become a symbol of the "sins" of the people living with them. People in better circumstances see laziness, crime, and pests, where the effects of racism, lack of opportunity, or simple poverty truly reside. When we dehumanize each other, we even use words associated with pests. The Nazis compared Jews not just to animals, but to rats. People kept in oppressed conditions are compared to vermin, and suffering from the depredations of those vermin is a by-product of their oppression.

We fail to manage pests in all sorts of ways. But sometimes the reason we fail is because we also fail to stand up for each other. Human areas become infested because we haven't given everyone the dignity of clean, well-built, vermin-free homes. We don't want to spend the money to put systems in place to help people fight off animals they cannot control themselves. Another species succeeds because our social contracts have failed.

Humans and wildlife will only continue to run into each other as we expand out over the planet. As we take up more and more land and turn it into subdivisions, strip malls, and farms, there will be less space for the animals that once lived there. Squeezed into our parks and our backyards, their choice is to adjust or die.

When they die, we wring our hands. We mourn the loss of mountain lions, eagles, and pandas. We set up conservation schemes and celebrate any small success. We even keep track of their sex lives. When zoos closed during the COVID-19 pandemic in 2020, two pandas living in Hong Kong finally had sex—and people around the world shared the puerile panda pics.

But when animals adapt well (and maybe occasion fewer accidentally racy photo shoots), we don't seem to be so thrilled. What about when a raccoon learns a disused chimney is a warm place to raise a family, when a black bear learns that campsites mean food, and when an elephant realizes the fields outside a national park are there for the taking?

In theory, we wanted these animals to live. We just didn't want them to live *here*.

TO MURDER SHAKESPEARE: Some are born pests, some achieve pestdom, and some have "pest" thrust upon 'em. But these are labels that we project onto these animals. Pests—the mice, raccoons, and seagulls of the world—are not irritating by nature. Instead, they are animal winners on a planet full of loss. When your habitat is full of parking lots, brick apartment buildings, and carefully tended gardens, survival isn't about staying sweetly in the woods and meadows. Instead, evolutionary success looks a lot like raiding our trash, nesting on our buildings, and eating our gardens down to nubs.

My friends take pictures of squirrels and send them to my group chats. My neighbors catch F***ing Kevin in the act. It makes us all smile. It's a funny, lighthearted mockery of my own failure. My personal squirrel menace is a symbol of my inability to put up nets or cover my plants, to create a less squirrel-friendly environment. Maybe he's a symbol of my failure to grow a garden at all.

But if I change the frame of this story, Kevin is something else. He's a story of success. I've created an environment that is perfect for Kevin. There's birdseed. Big trees for a squirrel to nest in. There are plenty of tomatoes. And, unless you count the cars, there are very few predators to trouble him.

Is it any wonder that Kevin wanted in?

F**KING KEVIN, MUCH as I hate him, has inspired me. He's made me wonder: Why does someone like me—someone raised predominantly in a Westernized society, living an urban and suburban lifestyle, someone who feels herself to be educated and open-minded, and who loves the environment so much—see some animals as pests and others as not? Why do some animals succeed so spectacularly at living with us, while others fail? And what do our pests tell us about our impact on the world?

I dug into history, research, religion, and rat nests. I spoke with scientists, historians, wildlife managers, pest control operators, and philosophers. I also spoke with Indigenous people—people who grew up with traditions very different from my own. Where I could, I spoke with members of Indigenous groups directly. Sometimes, though, I could not. Western views like mine—including the previous work of Western science—have done harm to Indigenous people and their land, sometimes pushing them off it entirely and often stripping them of the ability to protect their ecosystems and to preserve the knowledge they gained over so many generations. Journalists, also, are not always welcome, as previous coverage has not been fair or even polite.

When Indigenous people were generous and spoke with me, I listened and learned, but I know that no one person, or ten, can speak for everyone. Every group contains multitudes and many different ways of looking at the world. But we are all of us influenced by the culture we live in, and when we find ourselves in opposition to another creature, it's often our culture that determines whether we see a nonhuman neighbor—or a pest.

In addition to learning and listening, I went to the pests themselves. I hunted pythons in the dead of night, ate pigeon, and went on a rat safari. I tamed feral cats, tracked down drugged-up black bears, and reached into a coyote's rotting stomach. I smelled elephant repellant and shared a pigeon's lice. I did it because I wanted to know just why we hate some animals so terribly much— and why we love others, despite the harm they may cause.

What is the source of our disgust, our fear, and our disdain? What makes some animals so much better at living alongside us than others? How do our beliefs make us overlook animal harms? Why do some animals make us feel so powerless? And why, when we rip down their habitats, do some animals manage to persist?

I've chosen vertebrates—birds, reptiles, mammals like F**ing Kevin—as the subjects of this pest tour. To many Westernized people, bugs aren't as complicated as rats, cats, or deer. Insects don't cause many of us feelings of internal conflict. Most people will smoosh the bugs they hate (roaches, centipedes, take your pick) with a shoe and a small sense of triumph.

Vertebrate pests are another matter. We can see them in more than one way. We can put poison out for rats and protest their use as laboratory animals. We can shoot deer in the fall and show their adorable offspring to our children in the spring. Vertebrate pests lay bare our internal hypocrisy—how the natural world fills people who live separate from it with both adoration and dismay. Maybe that's a good thing. Maybe we need an animal reminder now and then to keep us humble.

I hope that, like me, you'll end up with respect instead of disdain. Awed by sheer adaptability and persistence, I hope you'll even cheer for a rat or two. Pests are proof not that nature is out to get us, but that it's all around us. Nature lives in our walls, poops on our heads, and eats all our tomatoes. Pests are what happens when we think we've got nature all figured out, and nature decides to give us the finger. Their story is one of human irritation—but also of animal triumph.

PART I

FEAR AND LOATHING

1

A PLAGUE
OF RATS

n the town of Deshnoke, India, a nine-hour drive from New
Delhi, lies the Karni Mata Temple. Its magnificent silver doors
sit in an intricately carved facade of white marble. Both the
doors and the carvings feature the beings that are venerated
inside—twenty-five thousand black rats (*Rattus rattus*). Or at least
there might be twenty-five thousand. There are more than anyone
can count, anyway.

The inside of the temple features a beautiful marble floor for
the rats. The rats drink from large bowls of milk, and a special
kitchen is devoted to their meals—fruit in the morning, brown
bread in the afternoon, sprouted beans in the evening. Devotees
share the rats' meals without a qualm. The rats dart everywhere. In
and out of crevices, across the floor, and along railings. They scurry
past groups of worshippers and sacred flames lit as people pray to
Karni Mata—the goddess of the temple. They poop everywhere.

Devotees (and a lot of tourists) come in droves, making food and fire offerings that the rats receive. In a requirement that might give people extreme cold feet, you have to enter the temple barefoot. That's so you won't hurt a rat if you step on it.

Karni Pratap, a medical undergraduate, goes to the temple to make offerings daily. "I've been a devotee since I was born," he says. "My family has been visiting the temple for generations. As soon as I grew up . . . I was visiting every day." In fact, when Pratap became ill as a child, his mother asked the goddess Karni Mata to cure him. When he got well, his name was changed to Karni—to the glory of the goddess.

To Pratap, and to other devotees, these rats are not actually rats. They are people, reincarnated as rats by the goddess.

Karni Mata was a human woman, born in the fourteenth century. As she grew into adulthood she became known as a sage. When one of her sister's children drowned while playing, her sister brought the boy's body to Karni Mata and begged her to bring him back. Karni Mata prayed to Yama, the god of death. Some say she fought the god and some she pled for his mercy, but either way, she won. Karni Mata said that none of her male family members would ever see Yama when they die. Instead, they would become rats, and when those rats die, they would rejoin the family. "These are not basic rats," Pratap says. "They are like a human incarnated in a rat."

The descendants of that family still worship at Karni Mata. Pratap is not related, but he worships at the temple every day in the hopes that he might be made into a rat when he dies. Sometimes, he says, the goddess has compassion on someone who is devoted enough.

Pratap has watched the rats closely since he was a child. "All the behavior you see in these rats is extremely similar to [that of] human beings," he says. "They eat like humans, or they're angry. Sometimes they're sad. Sometimes they play with each other, they fight with each other."

A certain amount of this seems like myth. Pratap says the rats don't breed too much, so there are never more than twenty-five thousand (since the rats don't get counted, no one can say if that's true). He says they don't leave the temple at all (which might be true, as rat ranges can be very small), and that when they die they don't smell, and that the temple itself does not smell.

Tourists from all over the world go to Karni Mata. Some are there to worship the rats. Others, to test their courage. Videos show people jumping, shrieking, and daring each other to touch a rat or let one crawl on them.

Pratap is happy to let the rats of the temple climb on him (they have to choose you, he stresses; you should not just pick one up). I ask if the rats of the temple changed his view of other rats.

"It's a feeling of disgust when you see a rat," he says. "Even in your house . . . [even] me being a devotee of this temple, if a normal rat comes to my house, I also scream and jump. I'm so scared of rats." Rats in his house? Horrible. Rats in the temple? Those are fine, he believes, because they are not real rats.

What's different is not the rats themselves. They're probably both *Rattus rattus*. A rat in the kitchen is a sign of disorder. It's the embodiment of dirt and illness, scurrying into your home. It's gross and even evil. A rat in a temple, in nature, or in your house as a pet? That's a very different matter. That's a rat out of pest context. It's a rat in a clean place, sometimes a holy place. It's clever, humanlike, sweet and even cuddly.

THERE ARE WELL more than fifty species of rats, but not all rats are created equal. Taxonomists have attempted to straighten this out in various ways. They tossed 167 species into the "tribe" Rattini (please, someone develop a cocktail called the Rattini), which includes a whole bunch of species in Southeast Asia, some of which are rats and some of which are not. A tribe is what taxonomists put between a family and a genus when they have no idea what's going on. The problem is that, for a long time, nontaxonomists have labeled pretty much any scurrying, tiny, hairy mammal with a bare-looking tail a rat, from the pack rats and kangaroo rats of North America (genus *Neotoma* and *Dipodomys*) to the greater bandicoot rat (genus *Bandicota*) on the Indian subcontinent.

But only two species have become the global species that we probably picture when we think of a rat—*Rattus rattus*, the black rat, and *Rattus norvegicus*, the brown rat. However, the black rat is not always black and the brown rat is not always brown. They both can be black, brown, gray, slightly ginger, and can even have nice spots. A few are white. On average, the brown rat is larger than the black rat, but there's a lot of variation. Brown rats tend to be geotropic—attracted to lower spaces like sewers or basements—while black rats are also called roof rats—attracted to roofs, trees, and higher spaces. But you can find a black rat low and a brown rat high.

Brown rats tend to outcompete black rats at colder temperatures. Probably only brown rats live in New York City, for example. (This could say something about the rats or something about New York. Either way it's not flattering.) But until the brown rat arrived to kick it out, black rats used to love chilly cities like London and still hang out in the British Isles, mostly in ports, alongside brown rats, or on remote Scottish islands. Both species exist in places like Los Angeles or Seattle.

Rat species can be so similar that when very picky zooarchaeologists see an ancient rat skeleton, they can tell for sure which species it is only if they have the entire skull or a DNA sample to work with. Honestly, the best way to tell with a live one is to take your rat's tail and bring it up and over their back. If the tail stops in the middle of the head, it's a brown rat. If the tail keeps going out past the nose, it's a black rat.

I understand if you do not wish to test this for yourself.

It often seems as though wherever people are, rats are too. But in fact that wasn't true until about five thousand years ago. That was when the black rat started spreading west from India, explains Emily Puckett, a phylogeographer—a person who studies the relationships between closely related species and their locations—at the University of Memphis in Tennessee. It didn't make it east from India at first, though, because it ran into another rat, *Rattus tanezumi* (which is the Asian house rat, which might also just be *Rattus rattus* with browner fur).

The black rat fared very well in the West, heading from India to Mesopotamia and the Mediterranean, then hitching rides with Romans, who went everywhere else. They arrived in Britain, where David Orton, a zooarchaeologist at the University of York in England, gets to dig some of them up.

When the Romans left Britain, though, rats were left in the lurch. Large cities, like Londinium, the capital of Roman Britain, were abandoned or reduced, but the lack of five-star restaurants isn't why the rats fled, Orton says. "It's not as simple as saying you need cities to have rats," he explains. Instead, what rats need is a system. They need infrastructure. Networks that take them from place to place and keep food flowing in and out. When the Romans left, Orton explains, they took the maintained infrastructure with them. The rat system collapsed.

The networks—and black rats—came back to Britain when the Vikings invaded and political and trade traffic increased in the ninth century. The species as a whole spent nearly a millennium as the undisputed rat ruler of Europe. "It is kind of the conventional view that until the eighteenth century, there are no brown rats at all," Orton says.

Black rats were from southern India. Brown rats "come from somewhere in the east, probably in northern China. They're also found living wild in Mongolia, and in parts of Eastern Siberia. Exactly where the starting point is, we don't really know," Orton explains. Colonialism brought European ships around the globe. They came back with new ideas, new materials, and new rats.

The nineteenth century also brought new ideas about those rats. People did not enjoy rats, of course—since they got in people's grain and ate crops. But oddly, people didn't think they spread disease. "The rat is thought of as having one redeeming characteristic, which is that it has no diseases and spreads no diseases," says Christos Lynteris, a medical anthropologist at the University of St. Andrews in Scotland. "I know it's shocking, but this is reality."

Rats were an economic problem—they were thieves of food. Huge numbers of rats might be seen as a harbinger of disease, but they weren't suspected of directly causing disease themselves. Even when writers described rats living in gross places, they often noted that the animals themselves were clean.

The rat's image got a makeover for the worse, Jonathan Burt notes in his book *Rat*, when underground sewers began to proliferate. Brown rats especially love sewers, and there's no amount of personal cleanliness that will take away the image of a rat coming out of a sewer covered in who knows what. The rat was officially disgusting.

Sickness—especially nausea—is what lies at the heart of disgust, explains Paul Rozin, a psychologist at the University of Pennsylvania in Philadelphia. Disgust is thought to be one of the basic emotions (along with anger, fear, joy, and so on) that are recognized across cultures. People confronted with disgusting material wrinkle their noses. The corners of the mouth turn down, and sometimes people stick their tongues out. Exclamations of "ew," "ugh," or "blech" are not uncommon.

These expressions are about "oral rejection," Rozin says. This makes sense, of course, when facing a potentially bacteria-laden food—like that takeout that you probably should have thrown out a week ago and which appears to be in the process of creating its own civilization. The facial expression says "Get it away from me."

With the tongue stuck out, it even says "Get it out!" "Guarding the mouth is very important, keeping bad stuff out," Rozin notes. "Disgust is the emotion that does that."

But the desire to get something out of our mouths doesn't perfectly explain our reactions to a rat. Hopefully, if you find a rat gross, it's never going anywhere near your mouth. So why be disgusted? Rozin specifies that it's because disgust as an emotion has achieved an expanded purpose over time. Somehow, roaches, mice, rats, pigeons, and so on got associated with things that were disgusting. Over time, the grime and disgust rubbed off.

Now, lots of things trigger disgust. You can feel disgust when you encounter poop, pus, or vomit. Bad food can be disgusting. You can be disgusted when you see a bug or a rat, and when you put your hand in something suspiciously squishy. You can even feel disgusted by things you find morally repugnant—like incest, racism, or cannibalism. Unfortunately, people can also report disgust when faced with people they have been taught to hate because of their gender, their race, or their religion—taking prejudice and making it visceral. "Disgust in and of itself might really be more than one thing," says Camilla Nord, a cognitive neuroscientist at the University of Cambridge in England. "And I think the best way to describe that is physiologically."

Some scientists like to look at emotions with measures like skin conductance—basically, how much does a particular feeling make you break out in a sweat. But that doesn't distinguish well between emotions, Nord says. "It's just sort of like arousal." So to study disgust, Nord went to the thing that disgust is linked to at its very base—nausea.

The emotion of disgust gets into your gut. "You can see differences in the rhythm of our stomach," Nord says. "So it goes from this three cycles per minute, quite slow, quite regular rhythm to either a bit faster or a bit slower." People will also glance away from the disgusting thing they are looking at.

Nord has shown that by altering this stomach rhythm, she can give people a higher tolerance for disgust. By giving twenty-five participants an antinausea drug called domperidone to chill out their stomachs, she and her colleagues were able to make them stare at disgusting stimuli longer—especially if they got paid.

Nord has not yet studied whether this gut feeling applies to moral disgust—or to disgust of rats. But here is a physical reaction that could play a strong cultural role. "My theory is that disgust actually evolved, at least in part, as a social emotion, something that people use to regulate social boundaries," says Joshua Rottman, a moral psychologist at Franklin & Marshall College in Pennsylvania. We might be disgusted by another culture's foods, he says, because we are distinguishing ourselves as not of that culture.

It can also be used to help people have a better reputation within their society. In that way, Rottman says, we might see disgust of rats as a way to enforce a social boundary. If disgust, he says, could make you more popular, more like other people you admire, "then it could make sense for disgust to kind of mark those things that we don't want to touch." If rats aren't seen as normative, if they're seen as, for example, signs of dirt, signs of poor home maintenance, then being disgusted by a rat is a way to say "I'm not like that."

THE SCIENTISTS QUICKLY decide to call this one Mariah Carey. The reason is abundantly obvious. Carey is known for her five-octave range, and this small female rat would give her sky-high voice a run for its money. It has squeaked, squealed, and downright screamed at an earsplitting pitch ever since Niamh Quinn, who studies human-wildlife interactions at University of California Agriculture and Natural Resources Division in Irvine, shook her out of a live trap into a thick cloth bag and got a grip on the loose skin of her back.

Mariah is a small black rat, captured only about one hundred

Handwritten marginal notes:

People do see invasive species with disgust

Plants like wineberries & for me, wisteria ↓ but why? It's beautiful. It smells wonderful but it kills my beloved trees.

yards outside Quinn's office at the South Coast Research and Extension Center in Irvine, in the foliage of one of their demonstration landscapes, installations of several tiny buildings in a row. Each has a pretty garden, exhibiting low, middle, and high water use to urge homeowners to plant up their yards with foliage that is less thirsty.

They might be even more motivated to follow the low water use example if they find out that all three rats in today's catch were from the high water use landscape.

The rats who make their homes here have a fabulous Southern California lifestyle. The Research and Extension Center has acres of orange trees and avocados. A smaller garden nearby boasts different varieties of tomatoes and peppers, all cared for by a set of master gardeners.

Such a healthy diet makes for some very fit rats. Mariah may be small, but her lungs are strong. Her feet are too. She writhes, wriggles, and kicks, acting as though Quinn and her colleague, technician Carolyn Day, have put her on their dinner menu, although they are only trying to attach a small radio collar. Mariah is having none of it and kicks the collar off again and again, squealing the entire time. After a number of attempts, Day manages to get the tiny collar around Mariah's even tinier neck. Weighed, tagged, checked for a vaginal plug (to find out if she might be getting ready to make more rats), and now fitted with her new jewelry, Mariah is placed back in her trap and then released on the sidewalk. She vanishes into the undergrowth—and mercifully stops squealing.

Perhaps her screams served as a warning to the other two black rats. With wild wriggles and mighty leaps, they both escape Quinn's sure but tired hands and disappear into the bushes—hopefully to be recaught the next day.

This is one of the control sites for Quinn's study—sites where, beyond little radio collars, she won't be interfering in any ratty lives. On other sites, Quinn is deploying snap traps and, in some cases, the poison difethialone. Difethialone is a second-generation anticoagulant rodenticide (or SGAR). It inhibits an enzyme required for the cycling of vitamin K, which is needed for creating the proteins that make blood clot. Without vitamin K, small bumps and bruises become massive internal bleeding. Death is slow and painful.

In 2020, California banned the use of SGARs for rodent control (with exceptions for areas around agricultural buildings) until scientists like Quinn have conducted studies to find out whether, and to what extent, difethialone and other SGARs also affect the native wildlife.

The effects of anticoagulants aren't limited to rodents (in fact, the lifesaving anticoagulant warfarin was first developed as a rat poison. It is still used today, though in some areas rats have evolved resistance, and scientists had to develop the more hardcore second-generation poisons like difethialone). Coyotes, birds, and other species that might eat rats loaded with anticoagulant bait can end up being accidentally poisoned. Other species, such as ground squirrels or pet dogs (and sometimes, sadly, humans), can also eat the deadly bait. SGARs might get rid of the rats, but they might inadvertently get rid of many other species too.

Mariah is part of Quinn's study to look at how to control rats with and without rodenticides. Some areas—like Mariah's—get no intervention at all. Others have classic snap traps and poison, and a third group have what Quinn calls a "mixed-management" approach, starting with snap traps and doing residual cleanup with difethialone.

Southern California is askitter with rats. The warm, dry environment hosts both *Rattus rattus* and *Rattus norvegicus* (and, Quinn suspects, *Rattus tanezumi*). But getting permits to do research on different sites is tough. "I'm still waiting to hear from a really ratty hotel" near Disneyland, Quinn says.

Whether people are coating landscapes in poisons or not, the rats can't stay. They freak people out, yes, but part of the reason people are so disgusted by them is their association with disease.

The disease Quinn is most worried about is called leptospirosis. It can be deadly to both animals and people—especially by the time it's diagnosed. Leptospirosis is a bacterial infection (from the bacterial genus *Leptospira*) that spreads through animal urine—which means that anyone or anything exposed to urine might be at risk. Often that's people working in slaughterhouses or sewers, although the disease is extremely rare in humans. What people in California are worried about is their dogs. Dogs love to sniff pee, of course. And a lot of rats leave a lot of pee.

Los Angeles County has seen an increase in canine cases of leptospirosis (although because labs were only required to report cases starting in 2014, the increase between 2016 and 2020 could just be reporting bias). But in 2021, an outbreak caused 201 cases of leptospirosis in dogs—and 13 deaths. There is a vaccine against leptospirosis for dogs, but uptake has been low, Quinn says, potentially due to expense and, during the COVID-19 pandemic, fewer vet visits. Most of the dogs got exposed to leptospirosis through wildlife, and the most common animals they interacted with were raccoons, rats, and mice. Leptospirosis is only one disease rats carry; since rats live around our waste it means they also carry all the bacteria found in it—like salmonella and *E. coli*.

People jumping on chairs and screaming when they see a rat, though, probably aren't thinking about leptospirosis. The first scream is probably just because they're surprised. But if pressed, almost everyone can tell you what disease we should really fear from rats.

Plague.

You might think that plague—the bubonic plague, the disease caused by the bacteria *Yersinia pestis* and famous for its swelling "buboes" in the armpits and groin and for killing off millions in the Middle Ages—belongs in the past. It doesn't. In fact, we are currently living in the third plague pandemic. The first you might have heard of as the Plague of Justinian. The second was the famous

Black Death. The third? That's now. It raced around the world in the late nineteenth century, caused millions of deaths in India, China, and Hong Kong, and spread to the United States. There, it added panic to a potent stew of racism and xenophobia over Chinese immigration and contributed to the permanent extension of the Chinese Exclusion Act. Y. *pestis* now makes a home in the United States in rodents like ground squirrels and prairie dogs. But it's not over. Every year, people in places like Madagascar, Mongolia, and the Unites States get plague.

Plague is not a human disease, not really. It's a rodent disease and a flea disease that sometimes happens to boil over into humans with deadly effect, explains Joe Hinnebusch, a medical microbiologist and chief of the plague section at the National Institute of Allergy and Infectious Diseases in Hamilton, Montana.

When a flea encounters a host with plague and bites it, it slurps up Y. *pestis* along with the blood. If the flea is susceptible to the bacteria, Y. *pestis* will form a biofilm across the inside of the flea's upper gastrointestinal tract. This effectively seals off the flea's stomach. The ravenous flea bites, again and again, but just vomits right back up any blood that it drinks. The flea starves to death.

You'd feel bad for the poor hungry flea, until you realize that every time the starving thing chomps back down, it barfs up bits of its bacterial film along with the blood it tried to eat, spreading the plague around to whomever it bites.

Most of the time, the fleas are biting rodents, and often those rodents are rats, though they can also be squirrels, marmots, chipmunks, guinea pigs, prairie dogs, mice, and even rabbits. These animals also get sick and die—and their fleas hop onto other nearby rodents. And so the cycle continues.

Humans encountered plague long before we ever built cities full of rats. Scientists have found Y. *pestis* in a 5,000-year-old skeleton from Latvia and a 3,800-year-old skeleton from Russia. The genetic evidence from the bacteria indicated that the Russian skeleton could have gotten plague from a flea. But the 5,000-year-old skeleton told a different plague story.

There's more than one way for a person to get plague—technically there are three. They could be bitten by a rat flea, a human flea (yes, many species have their own, highly discriminating fleas) or a human louse. Someone very unfortunate can end up with pneumonic plague in the lung, and spread it around directly through their infected tissue or hacking cough. This last version is by far the most terrifying, but also the most rare—because it kills its victims so quickly that they don't have time to go around coughing on that many people. This method—or a method directly from the bite of an infected rodent—may have been the version that got the 5,000-year-old Latvian skeleton.

Which version of these caused great pandemics like the Black Death? No one actually knows, and historians and scientists get into massive fights about this—the kind of passionate debate that is possible when everyone directly concerned with the issue has been dead for a few hundred years. But a lot of people are trying to find out. For example, Katharine Dean, a researcher at Norwegian Veterinary Institute in Ås, compared different models of plague transmission, mapping them onto death rates in different locations during the Black Death. In her findings, the best model was the one where human ectoparasites—fleas and lice—were the main drivers.

But while plague bacteria can easily plug the gut of a rat flea, they don't easily plug up a human flea, so human fleas transmit plague inefficiently. So inefficiently, Hinnebusch says, that it would require far too many human fleas to pass on plague to their human host. Humans would have to be weighed down by a veritable blanket of fleas—one they certainly could not ignore as they tried to go about their daily business.

Dean says she's open to alternative interpretations. She also notes that other scientists have shown that lice can carry plague as well, and lice can hide in things like the seams of clothing.

Ben Bolker, a mathematical biologist at McMaster University in Ontario, is not so sure. "I'm mildly skeptical of the whole ecto-parasite story," he says. In Bolker's simulations, he's created models with different likelihoods in different locations, based on better data from the third pandemic of the plague. Some of his cities end up with human ectoparasites, some with rat fleas. Some are pneumonic. "I don't think we're really ever going to know," he says. Academics can keep bickering forever.

While the people experiencing the Black Death didn't make the connection between rats and fleas and plague, they certainly killed rats. They also killed cats, dogs, and any people they thought might have poisoned wells or might have brought the disease to their cities and towns. But when scientists finally isolated the bacteria (Alexandre Yersin got the name for isolating *Yersinia pestis* in 1894, but Kitasato Shibasaburō also isolated the bacteria that same year) and connected the dots (Paul-Louis Simond gets credit for associating plague with infected rat fleas in 1898, though Masanori Ogata also made the link in 1897), the war on the rat began in earnest. It was now an officially pathogenic animal.

In Hanoi, Vietnam, French colonial administrators put on an official rat hunt in 1902, offering bounties for rat tails. To their consternation, the colonizers found that the Vietnamese knew a good opportunity when they saw one. They started breeding rats in the country to send tails to the city, and French officials even saw tailless rats running the streets of Hanoi, set free again to make more rats—and more money.

On the island of Java, the rat hunt took the form of a building project. "The Dutch colonial doctors became obsessed with bamboo as a conduit of rats," Christos Lynteris says. At the time, Indigenous people's houses were built of bamboo, and as Dutch doctors found rats living inside the bamboo, they quickly decided that the houses themselves were plague-ridden, and that Javanese houses—and people—needed to be sanitized. So the Dutch had the villages destroyed. They razed more than a million homes. The Javanese were required to build with wood instead. This put a huge burden on the local people. Not only did they have to tear down and re-build their houses, the Javanese could no longer do repairs with local materials—because the local material was bamboo. They had to buy timber. "You don't have money for timber, so you have to get in debt," Lynteris says. "It's basically an entire system of debt slav-ery." The house destruction was entirely in vain, and rats continued to afflict the people of Java long after the Dutch decamped. All in the name of getting rid of the disgusting, disease-causing rat.

ASSOCIATING RATS WITH disease in urban environments might help us avoid getting sick. But when we transfer that disgust to all kinds of rats in all places, all the time, it also cuts us off from what else they could be, and what else we could see.

When the ancestors—the tupuna—of the Māori first came to Aotearoa/New Zealand (Aotearoa is the Māori name for New Zea-land) around the thirteenth and fourteenth centuries, they came prepared with a long history of sea voyaging, which means they brought the animals and plants they would need with them. They carried foods such as sweet potatoes, coconuts, and breadfruit in their boats. They also packed the Pacific rat, *Rattus exulans*, which they call kiore.

Kiore were not gods or spirits or pets. They were food, usually for special occasions. The animals lived all over the island. Each tribe (iwi) had access to particular areas where they could trap kiore in the right season. Traditionally, notes Brad Haami, an author who specializes in Māori knowledge systems and is part of the iwi Ngāti Awa, kiore were trapped, skinned, and roasted. The fat got saved and set aside. Then the roasted kiore was put in a hollow gourd and the melted fat poured over it, which acted as a preservative.

Haami has eaten it, and so has Mere Roberts, an ecologist who identifies as having both European (Pākehā) and Māori (Tainui) descent. She was studying kiore for her dissertation. "The unofficial rule in my department of zoology in those days was that you ate your study animal," she says. "So we did on this island of Tiritiri Matangi. We cooked up some kiore and ate them. I must confess, we cooked them in lashings of red wine. And we called it ratatouille, appropriately. And it was very tasty."

(Māori are far from the only group that has rats on the menu. Rats are eaten in China, the Philippines, Thailand, and many other countries throughout the world. I admit, I really wanted to try one for myself, but a global pandemic prevented an epicurean adventure to Thailand.)

The kiore are food, yes, but like all the things that Māori people saw and named and studied on Aotearoa/New Zealand, they were also distant relatives of the Māori, explains Haami. The Māori keep detailed genealogies (whakapapa), going all the way back to their creators, Ranginui and Papatūānuku, who produced a series of children. "Each of them are seen as siblings," Haami says. "And each of those siblings produce something from the ecological world." One is the ancestor of sweet potatoes, another of fish. One is the ancestor of humans, and one of kiore.

Kiore have a respected place in Māori tradition. Before colonization, when Māori planted gardens or harvested foods, they left part of their harvest as a buffet for the kiore. The other part was for the humans. There were pragmatic aspects too. "You do your best to make sure that they don't interfere," Haami says, with incantations to keep the kiore from the human half of the garden. Storehouses were built elevated, to keep rats away from the larder. Even a good and tasty neighbor needs a little management, however respectful.

Many people see rats—all rats—as outsiders, stowaways. To the Māori, kiore belong to the ecology of Aotearoa/New Zealand as much as the Māori do, says Hōri Parata, a tribal elder of Ngātiwai. "A kiore didn't steal his way here, he was brought here deliberately. The kiore is properly here alongside us."

But Western scientists consider the Pacific rat part of the pest canon, one of the many rats eating the ecosystems hollow on Aotearoa/New Zealand. They are on the list of animals set for eradication by the Predator Free 2050 strategy, which aims to rid Aotearoa/New Zealand of possums, ferrets, stoats, weasels, and rats.

The kiore hasn't been able to make it on mainland Aotearoa/New Zealand since brown and black rats arrived. It's much smaller, and now persists on offshore islands, where it lives on fruits and eggs. Two of those offshore islands—Mauitaha and Araara—are reserves for the kiore. But Māori have granted permission to have them exterminated on the other islands. "They were having an impact on island biota," Parata notes.

Parata fought hard to preserve the two islands where the kiore are allowed to flourish. Scientists continue to study them, assessing their body condition and hoping that the island populations might be managed to allow Māori to trap them again.

"Why is it important? It's like asking why is my language important," Parata says. Kiore are part of Māori life and culture. They were brought to Aotearoa/New Zealand as part of their traditional foods—something lots of people can relate to. What is fall without apples and pumpkins? Summer barbecues without hamburgers, watermelon, or corn? English breakfast without ham? Everyone has their tastes, but for many, food and culture are deeply intertwined. For Māori, what is a feast without sweet potatoes and kiore?

Many modern Māori, though, Parata says, are stuck in Pākehā ways of thinking. "A lot of them don't even know what a kiore looks like," he says. "That's how Western paradigm talks: A rat is a rat is a rat. We don't call [ours] a rat, ours is a kiore."

IT'S 5:30 P.M., and I'm in Thomas Paine Park, facing the New York State Supreme Court building. It's late fall, already dark, but the night is warm. As I approach, it's not hard to spot the man I'm looking for—the only one waiting with a hard hat and safety vest: Bobby Corrigan, who runs RMC Pest Management Consulting and is widely regarded as a minor deity of urban rodentology. He's waiting with Chris Ciuro, a supervisor in Integrated Pest Management for the City of New York, who often assists him in studying how best to deal with urban rats.

Corrigan, Ciuro, and I head to meet Marlies Taley and Carrie Baker, who will both be joining us on our "rat safari." They are waiting for us on the other side of the park. Baker is radiant with excitement. She's been obsessed with rats since she read Robert Sullivan's book *Rats: Observations on the History & Habitat of the City's Most Unwanted Inhabitants* (which, to be fair, would make most people a little fixated). Taley got her friend a rat safari with Corrigan for Baker's birthday, but COVID-19 has put it off for almost a year. Now the evening has finally arrived.

Corrigan warns everyone that the night is young, and rats can be shy. "It's very much like whale watching," he says. Except for the flashlights, the tall buildings, the constant honking of horns, and the piles of garbage, I see the comparison.

We begin in Thomas Paine Park, where formerly lush yew bushes that used to trail branches to the ground have been ruthlessly hacked back. The goal was to provide no place for a rat to hide. Rats love nice, thick shrubs that go down to the ground, Corrigan explains. Liriope, pachysandra, and ground ivy are also popular. "You know how when you see water rippling, you're like 'there's a fish!'" he says. "You see pachysandra rippling in a line, that's a rat."

The plants give rats one of the three things they need—harborage, or a place to hide. Under comfortable, ground-covering shrubs or growth, they'll dig burrows about six feet long. One main entrance, a nice nest, and a back door, just in case. In Thomas Paine Park, the chopped-back plants appear to have worked. There are no fresh holes in the ground that we can see.

Rats also need water and food. Water they have in abundance. Lower Manhattan is built on top of a maze of streams and ponds, Corrigan explains, which have essentially been paved over. There are also plentiful storm drains and the many beverages people throw away. The food, of course, is everywhere too, in the form of open garbage cans, tossed bagels, and bags and bags of trash.

Sure enough, in a park on Canal Street, we spot a rat. It looks like it's been through the wringer, though, with cuts all over its back and a clearly broken and bleeding tail. Ciuro explains that he's been treating this park for rats for months, pumping carbon dioxide into burrows and dropping in poisoned bait. But they keep coming back. And when they do, the new residents fight for dominance.

The park is a maze of rat burrows. Hidden lightly under the long bushy grasses and ground cover, they're impossible to miss with the aid of a flashlight. "I just treated this!" Ciuro says. "I treated fourteen burrows right here!" He gestures to a raised planted bed. There are clearly more than fourteen burrows now.

Trash management is the key to rat problems, Corrigan says. Some of the parks have metal trash cans, enclosed in a diamond mesh—a perfect rat climbing wall. Businesses and residents alike put out trash in soft, easily chewable plastic bags the night before pickup, leaving a buffet available until dawn. Solid cans with lids would help make a difference, Corrigan notes. But people would probably hate the noise when garbage trucks came by. Similarly, he says, the city could make a rule that trash goes out in the morning, instead of the night before. But no one wants to get up earlier than they have to.

The parks could use better trash cans, like the Bigbelly cans. Huge, squat, and square, the trash cans have a drawer for a lid and are also a solar-powered trash compactor. The city has put in some

Bigbelly cans, but not nearly enough, Corrigan explains. Part of the reason is the hefty price tag: a single bin cost about four thousand dollars in 2019. The other reason is that compacted trash takes up less space. The cans don't need to be emptied as much. If you empty the cans less often, workers get paid less overtime, which the workers definitely don't want. People might not notice fewer rats. But they notice louder mornings, annoying ordinances, and smaller paychecks.

No matter how often it's picked up, trash does pile up in New York City—in some places more than others. "There are wealthy communities that are infested with rats, there are middle-class communities that are infested with rats," says Dawn Biehler, author of *Pests in the City.* "But a particular kind of infestation, where people are in very close contact with a lot of rats, comes out of the inequalities in housing investments that we see in American society."

Some of those inequalities are in housing that has been built badly and not maintained. But the biggest inequality in housing is between those with it and those without.

THE SKY IS leaden, and Heather Barr, a supervisor on the Health Education Action Resource Team (HEART) at Public Health—Seattle & King County, heard a rumor there could be snow. It doesn't smell like it, though. It's just cold and damp and depressing in the way only Seattle in December can be.

Barr and her team have brought me out to a camp they call Tenth and Dearborn, after the cross streets. A multitude of tents and tarp-covered shacks rises up the side of the hill and runs along the road, weaving between piles of thick, wicked brambles. In between the brambles, the ground is mostly mud, and most tents are on old pallets to keep them off the ground—and away from the rats, Barr notes. Barr's team is here to administer COVID-19 tests (which are in high demand) and to see if anyone wants a vaccine, which they can also give. They hand out harm-mitigation materials, from Narcan to clean crack pipes. They also distribute tents, sleeping bags, tarps, hats, gloves, socks, handwarmers, water, snacks, personal hygiene products, even a few precious donated cell phones—anything they can to make living outside just a tiny bit more comfortable.

The team parks and opens up the back of their city-issued van, pulling flats of water into a wagon and onto the sidewalk. The team's nurse, Dora Henricksen, gets her rapid tests ready. People start lining up, and Felicia Staley, one of the community health workers, pulls lightly on my arm.

"Did you see the rat?"

I turn quickly and see the last bit of a leaf ruffle where the rat dashed out from the brambles and under the chain-link fence next to the camp to race down the gutter.

A woman comes up with a tiny dog on a leash. Tiger looks like he could be part Yorkie and part Shih Tzu, but mostly he's friendly and carefully protected from the cold in a fleece doggie onesie. Tiger's owner, Alexis, has startlingly huge, beautiful eyes. But she is also pale and frazzled. There was a fire that burned down some of the tents the night before, and city workers are further down the block, tossing charred, soaked belongings into a truck. She asks about hats and gloves, as well as any chance there might be a room for the night, it's so cold.

As the team hands out supplies, there's movement behind the nearest tent. Another rat whisks out of a hole in the hillside and commences trotting back and forth behind the tent, nosing for food. It seems completely unconcerned by the people all around. It's the middle of the day. Rats shouldn't be active at this time, unless there's a very large population and food supply.

Alexis and her boyfriend have been living out here on and off for six years, she says. The rats are a constant problem. "They have no fear at all," she says. "It sounds horrible, but we use our BB gun and shoot them. It's actually kind of fun." The rats chewed a hole in the corner of their last tent. "We were putting cheese in the hole and trying to get them to come so we could shoot them." The two turned their backs for just a few seconds, a rat whisked in, took the cheese, and was gone.

"Once they get in your tent, they never go away," she says. "They'll kick you out." She and her boyfriend (and Tiger) had to abandon that tent.

The HEART team walks through the camp, calling politely into each tent to ask if anyone would like a COVID-19 test or a snack. Among the mud and trash, a dead rat lies curled against a beer can. Across the street at the top of the hill, the tents continue, marching down either side of a stairway. Fatima Guled, another of the community health workers, calls my attention to another rat dashing between the tents.

This isn't actually the rattiest camp, Staley says. In others, the hillsides are pockmarked with holes. Before she took a three-day "Rat Academy" course from Bobby Corrigan, she didn't realize what the holes were. "At first I was like, oh, there's a lot of snakes out here," she says. Now she knows what they really are, and the whole team is adept at spotting rat signs.

"A lot of times people bring it up to us, tell us there's a big rat problem," Guled says. "They ask 'what can you guys do about it?'" Barr and her team can't do much, they aren't pest management, though Barr hopes to get someone from environmental health on the team, which might help open more pest management possibilities. Right now, they try to help people living in encampments make do with what they have, handing out trash bags and showing them how to store food safely with the materials they have available.

A broom lies between some of the tents, flat in a cleared space where someone had just dropped it. "That tells you someone was trying to clean, right," Barr says. She wants to be able to give people experiencing homelessness empowerment and the tools to try to fight back the rats. "They start to feel very helpless when they can't get a toehold on things."

LIVING WITH RATS adds yet another stressor to people already heavily burdened, especially those living unhoused or in poor housing, says Kaylee Byers, the deputy director of the British Columbia node of the Canadian Wildlife Health Cooperative in Canada. Rats are stressful in themselves, but they're also a sign of a stressful living situation.

Byers has studied how people living in Downtown Eastside of Vancouver—an area known for poverty, large numbers of unhoused people, and drug use—deal with rats. Working with the Vancouver Area Network of Drug Users, Byers and her colleagues interviewed twenty people living in the neighborhood. Half of them described living with rats, either while they were unhoused or in their homes, and most said they saw rats daily.

The encounters people had with rats—especially people experiencing homelessness—had deep impacts. Not just mentally, but physically. "One of the things that was most vivid to me was people talking about the sounds made by rats," Byers says. People didn't just end up afraid, they ended up unable to sleep. It reminded me of all those movie scenes, where someone thinks they hear a rat, and sits up, clutching the bedclothes, only to find it's something innocuous. Only, in the Downtown Eastside, it really is a rat, and it doesn't go away when you lie back down.

The rats also contributed to the view the residents had that their neighborhood was less than, looked down on by other people in Vancouver, Byers says. Participants suspected that "if this was a different neighborhood, there would be more management of the rat issue." They felt the city didn't care about them, and that they didn't count.

The feeling of living in a run-down neighborhood, of being forced to live with rats, made residents feel angry, powerless, and stuck. "Your hands are tied 'cuz where're you gonna live, for one thing?" one resident said. "You're stuck here anyways in this shit, right?"

BACK ON THE rat safari, Corrigan and Ciuro walk swiftly along, but they both have a peculiar gait. When they walk past a pile of garbage bags, they step and deliver a soccer kick to a bag. Step, kick. Step, kick. They're trying to startle any rats inside, get them to go racing out from the banquet of garbage and into the storm grates where they live.

"This is the rattiest park in all of New York City," Corrigan says, as we stop at Columbus Park. "Was!" Ciuro replies. He explains with pride that he and his team treated every burrow in the park with carbon dioxide. After Ciuro and his team applied CO_2 for two or three months straight, he's prepared to swear there was not a rat left.

They used carbon dioxide instead of the usual poisons because there are also hawks nesting in the park. SGARs work—but often too well. When predators like hawks eat poisoned rats, they end up with a dose of poison themselves. And hawks in New York subsist mainly on rodents. So far, California and British Columbia have banned the SGARs for residential use. Corrigan and Ciuro are experimenting with other methods, like dry ice (called Rat Ice when used for rodent control for reasons only the EPA can answer for) and now CO_2 gas pumped from cylinders into burrows.

Columbus Park got the CO_2 treatment, but it hasn't been treated for more than a year, since COVID-19 gripped the city. I ask Ciuro if he's still confident there are no rats. After all, rats might have moved elsewhere in the early days of the pandemic, when all the restaurants were shut. But now, the places around the square are doing a brisk business. "Not at all confident," he says.

As we turn back to face the park, a big, healthy rat races right in front of us. Ciuro laughs.

Marlies Taley and Carrie Baker have been having a wonderful time. Baker peppers Corrigan with questions, and Taley is also into the spirit of the rat safari. We saunter around the park, excitedly pointing out rat signs: a burrow here, moving ground cover there, the dark, greasy stains marking the edges of walls—oil from rat fur as the animals scuttle by.

I keep asking Corrigan why the city never seemed to get anywhere with rats. They keep baiting, hanging tiny packets of poison from sewer grates, laying out metal mesh on the ground to prevent burrowing, pumping CO_2 into rat burrows or filling them with poisoned pellets. It barely makes a dent. The trash keeps piling up, and the rats keep piling on the trash. People seem to consider the rats as just one other gross thing about living in New York.

Corrigan shakes his head. "What it would take is a plague."

Plague. We've lived through something very like it in the form of COVID—but since it's a respiratory virus, little in a rat's life has changed. Even when restaurants closed, rats did just fine, delivery tastes just as good as dine in to them.

Corrigan is talking about a plague that can be blamed on rats. But I worry that even if that happened, cities would just try to bait, poison, and use CO_2 to suffocate their way out of the problem. What Corrigan keeps trying to tell cities—and what most cities refuse to face—is that the poison and gas won't work on their own. The rats will come back, again and again, until the humans change their waste management. Humans living with rats would need to change where they put their trash and how they built their buildings and sewers.

Recently, New York City has started a new program of trash

containers to hold sidewalk litter and the trash from regular trash baskets. The goal is a centralized place to store trash before it gets collected, to clean up the streets in places like Times Square. Hopefully, that will keep the iconic city's streets a little less littered and will keep the rats down.

Even then, there would still be rats—there would just be a lot fewer of them. Because wherever humans choose to congregate, some of them will end up on the bottom of the social order, bearing the brunt of people's indifference. They will get stuck with poor housing as the only housing they can afford—or no housing at all, making keeping their areas clean and unattractive to rats nearly impossible. People will keep up sanitation in some areas better than in others. Some people will choose to litter, some will choose not to clean up their environments. Collectively, they will let sewers decay, out of sight and out of mind. People need to improve not just their sanitation, but the way they maintain their cities and the way they treat their fellow humans. Otherwise, where humans fail, rats succeed.

Rats, I realize, are not in themselves disgusting. They just make their way where opportunity exists. In the forest, hunted by the Māori, they are food. In the Karni Mata temple, worshipped as ancestors, they are clean and holy. When rats are disgusting, it is because of where they live and why they live there. Rats are disgusting because humans are—because we generate trash and sewage, and because we are indifferent to each other's sufferings.

I kick every trash bag all the way home.

2

A SLITHER
OF SNAKES

t was late December 2007, and Michael Cove had an eight-foot Burmese python in his pants. Two pythons, actually, but the second was only six feet and barely counted.

Cove, the mammalogy research curator at the North Carolina Museum of Natural Sciences, had been backcountry hiking in Everglades National Park when he saw scales on the ground. "I saw this snake and I was like, 'Man, that thing's beautiful!' but I had three or four miles to go . . . and then I went 'Wait a second, this is an invasive species!'" he says. At the time, Cove was a zookeeper, so he went back to try to catch the snake.

Catching turned out to be the easy part. "I didn't think it through," he admits. "I threw my shirt over its head." He wanted to get something over the snake's head quickly so he didn't get bitten. But then, there he was. Shirtless. Holding a snake. "I was hopped up on adrenaline and I'm holding this thing," he says. "I'm shaking. I'm looking at its teeth. It's trying to get loose and bite the shit out of me and I was like, 'Okay, now what.'"

He might have been shirtless, but he wasn't pants-less. Yet. One-handed, Cove took off his jeans and tied the legs in knots to make a bag. He shoved the python in, slipped a bungee cord through the belt loops to keep it closed, and hiked on. Later, he came across a second, smaller snake, and added it to his makeshift bag.

Cove completed his planned camping trip with his new reptilian companions and stomped out of the marsh in his underwear the next day to turn in his backcountry permit. "I asked the woman at the station, 'Hey, do you guys ever find pythons?'" he says. She had a better question. "What happened to your damn pants?!" Cove's pants were lying behind him on the sidewalk, writhing with snakes.

Since then, the pythons have multiplied. Scientists estimate there to be tens of thousands of Burmese pythons (*Python bivittatus*) slithering through South Florida. No one is quite sure how they got there. People may have released pets when they got too big. A hurricane might have dumped a breeding facility's wares into the wetland. Maybe it was a bit of both. Regardless, Cove had definitely not found the first one. There were pythons in the Everglades by 1979. And those pythons had to eat.

Pythons, it turns out, eat quite a bit. From observations between 2003 and 2011, scientists associated the snake's rise with an 88 percent decrease in bobcats, a 99 percent decrease in raccoons and opossums, and a total absence of rabbits. The obliteration is on a scale scientists normally associate with introduced animals like rats or cats released onto islands.

But this voracious predator has scales. When it comes to Burmese pythons in the Everglades, no one is talking about using gentle live traps and adopting out what they catch. No one is trying to exclude them with fences, give them birth control, or teach them to eat something else. No. When people talk about getting rid of pythons in the Everglades, they talk about guns, daggers, and machetes. On private land and in some wildlife management areas, you don't even need a permit.

I've talked to many people about the issue of snakes in the Everglades, and I'm often struck by their anger. I'm also surprised at their fear. Most people I speak to have never been to the Everglades, but knowing there are pythons as long as twenty-three feet sliding around South Florida gives them the crawling heebie-jeebies. From there, it seems a very quick slide from "the pythons need to leave" to "the pythons must die."

Did he eat them in Eating Aliens.

And I wonder how much of that anger, how much of that hate, arises from the very fear people feel around snakes. Burmese pythons are, without doubt, an invasive species. They are certainly disrupting the ecosystem, eating species to local extirpation—though not to extinction. But they are doing no more than many cats, ferrets, stoats, foxes, and other mammals have done when taken from their homes by humans and placed in a new environment that suits them very well.

Sometimes, the animals we call pests get a little extra help in gaining a nefarious reputation. They get that help in the form of our fear. As a very wise teacher named Yoda once said, "Fear leads to anger. Anger leads to hate. Hate leads to suffering." The suffering, in this case, of the animals we detest. This way to the dark side.

MIKE COVE RAN into his pythons by chance. If you want to catch pythons on purpose, you need materials, strategy, and a good amount of luck.

It's a little after one in the morning, and I'm standing on a platform on the top of a 1998 Ford Expedition (with custom python-skin interior and a vanity license plate that says SNAKER) as we drive slowly along a levy in the Everglades. A clammy, humid wind is blowing, and I'm pretty sure every single insect in the Everglades has found its way onto my face. The next morning, I'll pull an entire mosquito out of one ear canal. At least, I hope it was the whole mosquito.

The hum of the engine is almost drowned by a chorus of frogs; the big burps of pig frogs, the rattle of cricket frogs, and the low squeaker-toy sound of Cuban tree frogs. Floodlights glare on either side of the vehicle, catching the winking emerald glitter of thousands of spiders and the dark forms of alligators swaggering through the brush.

It's positively magical.

"Stop!" Jayna Parsley Corns shouts from the other side of the platform. By day, Corns is a makeup artist and hair stylist. By night, she's a passionate Everglades Avenger, a member of a team dedicated to helping remove pythons from the Everglades.

Donna Kalil, "the python-hunting queen of South Florida" and founder of the Everglades Avenger Team, slams the brakes, and we brace. Then she backs up slowly, handheld lights trained as Corns guides her carefully back. There, on the side of the road, lies something that shines slightly too brightly, curves slightly too gracefully to be a stick. It's a nearly six-foot-long Burmese python, probably contemplating a road crossing.

Kalil hops out, steps into the brush, and comes up on the snake carefully from behind. It turns and she nabs it, so fast I'm glad I got it on video. Cove needed a shirt and his own jeans. Kalil needs only one hand. She picks up the snake by the head, lets it wrap itself tight around one leg, and waddles back to the car.

It's Corns's first python spotting of the night. We all pause for victory selfies.

Data logging ensues, and then the snake is popped into a sack, and the sack is stuffed in a locked box. In the morning, Kalil will kill the snake (approved methods include captive bolt guns, air guns, and decapitation) and turn it in to add to her tally.

Kalil is a python elimination contractor at the South Florida

Water Management District. But this week, she's also participating in the Florida Python Challenge, which she has entered every year since its inception in 2013. The 2021 prize? Ten thousand dollars for the most pythons caught. Given that you're lucky to catch one python a night and the whole challenge is only ten days long, numbers are never high. The previous year's winner caught eight. And while size does count, it's the number of pythons that wins the big money. Kalil says the prize is likely to go to whoever finds hatchlings. The baby snakes all come bursting out of the nest, so where you see one, you're likely to see more.

Hunt Pythons at night

We continue on, passing other hunters. Each time, Kalil stops to chat. She's trained many of these men (they are all men) herself, and they greet her with deep respect. But chatting isn't just common courtesy. Stopping for a few minutes after one hunting team passes gives time for more pythons to emerge. It's also a good source of gossip. There's complaints about having to stay awake until the event stations open at 8:00 a.m. People trade tips about where they've caught snakes and mourn their own catching slumps. Rumors swirl that someone found twenty hatchlings.

We caught only one python that night. But when the results are released a few weeks later, Kalil has again proven her prowess, winning the 2021 Python Challenge in the professional category with a total of nineteen snakes. She did, indeed, find hatchlings.

The 2021 Florida Python Challenge registered 600 hunters and removed 223 snakes. Sure, 223 is a lot of snakes. But it's only about one third of a snake per snake hunter. Most hunters never caught any at all. The number seems especially pathetic when you consider the tens of thousands of pythons that are slithering through the swamps. You'd think, with pythons so thick on the ground, people could net at least one per person. Or ten.

But spend one night out hunting snakes and you'll realize exactly why capture numbers are so low. Our floodlights illuminate swaths of grass and bushes on either side of the levy—about ten yards or so on each side. Past that, it's a thicket of cypress and vines, and lots and lots of water. The ground that is lit is covered in branches, and in the beam of the searchlights, all of them kind of look like snakes. Humans go into the water at night at their own risk.

Pythons are also "cryptic," which is to say they could make every military camouflage designer weep with envy. Their dappled dark-and-light pattern is terrifically hard to spot, especially when they don't move. Try to look for them with thermal imaging? Good luck. Scientists are working on that, but pythons are exothermic, and unless they've been basking in the sun to warm themselves, they don't stick out well from the ground around them. Right now, any Burmese python that's going to get caught has to come to the roadside, slithering far enough out of the trees for the telltale glimmer of their scales to give them away. And then, of course, someone has to be brave enough to catch them.

Not too many people are. A lot of people are scared of snakes. About half of adults surveyed in places like the United Kingdom have a healthy concern. In 2 to 3 percent of the population, that fear rises to the level of phobia (ophidiophobia, if you want a cool word to use at dinner parties). That's a lot of people who get nervous around slithering scales. But there are also people who love snakes, who keep them as pets. And still more who don't really care one way or the other.

And because so many people fear snakes, a certain number of scientists have taken to asking why. In particular, they want to know if our fear of snakes is innate or an evolved tendency to associate snakes with fear. These might sound like the same thing, but there's a subtle difference. If it's an evolved tendency, we'd still have to learn to be afraid of snakes, explains Vanessa LoBue, a psychologist at Rutgers University in New Jersey.

Touch a hot stove once, and you'll never do it again. Seeing the glowing stovetop immediately triggers a memory of your previous burn. This is classical conditioning—the technique that made Pavlov's salivating dogs so famous. It's learning a direct association between a signal (say, a snake) and a response (screaming and running away as you remember that the snake bit you before and you nearly died). If we learned all our fears this way—the fear of burning, the fear of snakes—then we should also learn them all just as easily. It should be as easy to develop a fear of snakes as it is to develop a fear of dogs because you were bitten once, LoBue says. "But that's not really the case with fear. Because there are things that we're more likely to be afraid of. We're more likely to be afraid of snakes and spiders and other things. And is it because we're classically conditioned to be afraid of snakes?"

Because it involves learning that association directly, LoBue says, classical conditioning isn't very likely. Most people who are afraid of snakes have never been bitten by one. But if there's an evolved tendency to associate snakes with fear, she notes, the learning that snakes are dangerous might come easily, much more easily than someone would learn to fear, say, a dog or cat.

In the innate theory, of course, we'd never have to learn in the first place. Every baby would be born knowing a snake when it saw one. It's a popular idea, LoBue notes, but she has found some evidence it's not true. LoBue has done most of her fear research in babies, and she's found that when presented with images or videos of snakes, babies will focus intently. "Babies will try to grab snakes from [the] screen," LoBue says, something they'd never do if they were born to fear snakes.

Even monkeys don't always back up the innate theory. Monkeys born and raised in the wild are certainly afraid of snakes—as they should be. They even give models of snakes and "sinuous" rubber tubes some side-eye. Raise a monkey in the lab, though, and they usually show only a little fear, and snaky tubing scares them not at all. But once a wild monkey nearby indicates that the snake is bad news, lab-reared monkeys quickly pick up on their nerves.

LoBue still thinks people respond to snakes differently, but she doesn't think it's about fear. "The more I researched, the more I saw that, you know, maybe things like snakes and spiders are spe-cial," LoBue says. "But what's special about them is that they're so weird, they're so different from the other animals that we're accus-tomed to."

From the time we are born, we are surrounded by humans, mammals, and birds. These animals all move in a specific way—they move on four legs or two, and they have gaits associated with four or two legs. "And snakes don't move that way," LoBue notes. "Spiders don't move that way. . . . I don't have data to back this up, but their movement is really anomalous. . . . I feel like their motion is somewhat unpredictable."

LoBue has shown that when given a snake video to focus on, ba-bies will stare. But staring isn't the same thing as fear. The strange, unfamiliar motion, LoBue guesses, might be what's making them look. "I don't think it's about snakeness or spiderness, I think it's about novelty."

That strangeness might get a boost from some of the shapes that more venomous snakes tend to boast, says Fabien Aubret, an evolutionary biologist at Centre National de la Recherche Scien-tifique in Moulis, France. In a study of three-to-eleven-year-old children, Aubret asked them to describe various animals (bunnies, dogs, snakes, or smiley faces with pointy teeth) as "nice" or "mean." When comparing a blank, roundheaded snake to a snake with a pointy head and zigzag pattern, the children were more likely to label the pointier snake as mean. The same went for snarling dogs and smiling faces with pointy teeth. "So we would probably say that there's probably an innate reaction to some shapes in the envi-ronment," he says. It's moving in a strange way, and it's covered in "mean" pointy patterns.

"There's certainly something visually stimulating about the way the patterns and . . . the snake moves," LoBue says. Snakes are just arresting to look at. Venomous ones can have bright colors that are supposed to advertise their dangerous status. Of course you can't stop staring.

The innate-fear theory also doesn't quite pass the common sense test. After all, if we all had an innate fear of snakes, wouldn't more of us be afraid? Sure, half a Western sample said they weren't comfortable with snakes. Probably more than that would briefly jump if they saw the sudden wriggle of a snake in the wild, just from pure startlement. But half isn't everyone. It means that, if the innate theory were true, somewhere along the line about half of us would have lost that much-vaunted fear of snakes. And since there are no programs to carefully decondition our fear of snakes, maybe we're just not born with it.

Instead, like lab-raised monkeys, it might be caution—caution that becomes fear from the fear that other people have of snakes. LoBue has also shown that babies stare even more at snakes when they are paired with the sound of someone speaking nonsense words in a fearful voice. When spying on parents and children at a zoo's reptile house, she and her colleagues showed that parents provided more negative information about a python than about a Komodo dragon. Parents joked that the python might eat little kids. The Komodo dragon, on the other hand, might kiss them.

Kids are fast learners. In a survey of preschoolers, LoBue showed that 64 percent already had some fear of snakes, and 65 percent were already nervous about spiders. The children were also more likely to provide negative information about them than about frogs or turtles.

What parents don't provide, Hollywood is more than glad to make up for. From *Indiana Jones and the Raiders of the Lost Ark* to *Harry Potter* to *Snakes on a Plane*, our media writhes with slithering villains. They add to any unsettling feelings a snake's movement might create, telling us all that these animals are born bad.

We get most of our information about animals in general from culture and the people around us—from snakes and bears to chameleons and birds of paradise. Many people will never encounter most of the creatures they see on Animal Planet in real life, so we have to rely on portrayals in the media to make up the difference. That media is also often from our own culture. It's worth noting that LoBue was looking specifically at the fears of Caucasian people.

Those fears have a basis in more than just American ideas about evolution and folklore. They have roots in religion too. When people raised in Christian contexts think about snakes, for example, many of them can easily draw on the Garden of Eden. There, Eve (and Lil Nas X) is tempted by a snake. Eve eats fruit from the Tree of Knowledge, and humanity as a whole is doomed.

Original sin is a good enough reason for people to hold a snake grudge, but God adds an extra punishment. "Because you have done this, cursed are you above all livestock, and above all wild animals! You will crawl on your belly and you will eat dust all the days of your life. And I will put enmity between you and the woman, and between your offspring and hers; he will crush your head, and you will strike his heel." Judeo-Christian snake hatred isn't biological, it's ordained from on high.

As Christianity spread, even formerly beloved snakes became signs of evil. As Emma Marris recounts in *Wild Souls: Freedom and Flourishing in the Non-Human World*, the little grass snake (*Natrix natrix*) used to be known as a little god of the underworld in Europe. In the winter, people in Lithuania and Latvia would let the little black snake with pretty cream patches behind its eyes hang out in their houses for warmth, and people in Baltic countries considered them the reborn spirits of people's ancestors. It was the "good spirit" of the house, write H. J. R. Lenders and Ingo Janssen in their history of views on snakes in northern Europe. Only when people converted to Christianity did the grass snake take on a more sinister aspect. A new religion changed the harmless little snake into the basilisk—the symbol of the Antichrist.

But it's just too simple to base all our fear and hate on a few lines in Genesis, notes Félix Landry Yuan, who studies ecological conservation at the University of Hong Kong. In fact, there are plenty of positive snake references in religion. "Snake worshipping is found throughout the world," he says. "You have the Mayans, and the pre-Mayans, you know, the Olmecs. You have the Indigenous groups of Australia that have the whole rainbow serpent myth. You have the ancient Mesopotamians, you have the ancient Egyptians. And then you even have some cultures in China and Japan that worship snake gods and snake deities, and these are all very ancient religions with ancient belief systems predating Christianity." In Europe, faith-based ideas about snakes haven't always been bad either. The grass snake is one example, but the ancient Greeks portrayed snakes on the rod of Asclepius, a god of healing and medicine, and two snakes twined around the caduceus carried by Hermes.

Landry Yuan and his colleague Palatty Allesh Sinu, an ecologist at the Central University of Kerala in India, are particularly interested in beliefs about snakes in places like India's southwest, where snakes—extremely deadly ones—are an everyday fact of life. There are almost three hundred species of snake in India, and about sixty of those are venomous. (The technical term for a bite from a venomous snake is "snakebite envenoming," which sounds like something clinics in California will offer to get rid of wrinkles any day now.) These include the spectacled cobra or Indian cobra (*Naja naja*), the central Asian cobra (*N. oxiana*), the monocellate cobra or monocled cobra (*N. kaouthia*), and the king cobra (*Ophiophagus hannah*), which some taxonomists would say isn't actually a cobra, not that you'll care once it bites you. There are also vipers, pit vipers, and kraits.

None of these snakes are out actively hunting humans. But India is a densely populated country for both humans and snakes. Scientists estimate that between 2000 and 2019, India averaged 58,000 deaths from snakebite per year. By contrast, the United

States recorded only 101 snakebite deaths between 1989 and 2018. Bites from venomous snakes in places like India are so common—and so deadly—that the World Health Organization declared them in 2017 a high priority neglected tropical disease.

If India were like the United States, you'd think that the numbers of snakebites and deaths would send the entire population out into the forest with hoes and machetes. But by and large, Sinu says, in Kerala, the state on the southwest tip of India where he lives and works, they don't. "In southern Kerala, the population is fifty-four percent Hindu and forty-six percent Christians and Muslims," he explains. "Different relationships exist between the people, the community, and their attitude towards the snakes." When Christians see a snake, he says, "their first attitude is to kill." But for people who follow Hinduism, Sinu says, "I think the fear component is far, far less." Snakes play a more respected role in their lives—and especially in their religion.

Kerala has a large number of what are called "sacred groves." These are usually natural spaces with shrines or temples in them, and can vary from only about 0.17 acres—a small lot—to more than 100 acres of forest. The sacred groves "serve as the gathering points for cultural and traditional ceremonies and rituals, but also as places of worship," Landry Yuan says. "The sacred groves are seen as an extension of a temple, but outdoors."

In many of these temples, people are worshipping traditional local deities such as Muthappan, though Hindu gods such as Vishnu have also been brought to the groves. They are also very definitely worshipping snakes. When Vishnu takes a nap, he does it on the coils of an infinite, multiheaded snake. People also worship nagas, serpent gods, on their own. "And so when you go to some of these temples, for example, you might see that there's a statue [of] some kind of deity that's worshipped, but then you'd also see a naga that's worshipped in association," Landry Yuan says.

Sinu and Landry Yuan conducted surveys outside of sacred

groves and found that more than 50 percent of people asked said they liked snakes—real, live ones. The people surveyed also noted that they would not harm snakes in the sacred groves—though outside of the groves, only 60 percent said they would definitely leave snakes alone. And of the three hundred people surveyed, only forty-seven people (about 16 percent) expressed fear.

Worship, though, doesn't have to mean love, Landry Yuan says. One group was truly devoted to the snakes in the groves, but another offered worship out of fear. Worship would keep the snakes away—away physically and away from their dreams. I leave an offering, you leave me and mine unbitten. I'd take that trade-off.

In some cases, people in India are far more tolerant of deadly snakes than almost anyone in the United States would be. King cobras sometimes seek shelter in people's houses during the monsoon season, Sinu explains. Video clips abound on the internet of people finding huge cobras in their kitchens. King cobras might not be as large as Burmese pythons—the cobra tops out at a measly eighteen feet to the python's twenty-three. Because of their bite, though, king cobras are far more dangerous than any python that's ever haunted the Everglades. But in the videos, most people don't scream or hide. They call some snake catchers, and often go back into the room with them to show where the snake is. In other snake rescue videos, people gather in a circle, cell phones out, or crowd in behind snake removal teams, coming far closer than some Americans, who would avoid the reptile house entirely. People might worship out of devotion, or they might worship out of fear. But with worship comes respect. And with respect comes coexistence.

IT WOULD BE easy to think of humans as being good at driving species extinct—purposefully or otherwise. With our history of wiping out species, from the dodo to Steller's sea cow, you'd think we'd be able to eradicate a snake or two without breaking a sweat. You would be wrong.

The brown tree snake (*Boiga irregularis*) can get up to ten feet long, but usually sticks to between three and five feet. It is slightly venomous, slender, and—as the name so obligingly notes—brown. It has golden, cat-pupiled eyes and a perpetually smiling expression. (I'll be honest, I think it's adorable.) It arrived on the island of Guam in the late 1940s, courtesy of the military. "The island from which they were thought to have originated also had large staging areas for military materials," explains Shane Siers, an ecologist living on Guam with the U.S. Department of Agriculture National Wildlife Research Center. "You've got jeeps and tanks and pallets of stuff sitting next to the jungle." A snake prowls around looking for food and finds a nice dark wheel well to hide in for the day. They wake up from their nap to find themselves bound for Guam.

The birds and lizards of Guam had never seen a snake before. There are no snake predators on the islands either, with the exception of some monitor lizards. The brown tree snake began to multiply. Within about forty years, scientists estimated there were between fifty and one hundred snakes per hectare (a hectare is about the size of two American football fields). Guam has about 5,930 hectares. At the height of the "snake eruption," Siers says, the rough estimate was between two and four million snakes on an island about the size of Chicago.

There were so many snakes that they killed pets, sickened babies, and almost killed off the local bats. A more daily irritation was that the snakes like heights—like the nicely strung power lines. On the sensitive power grid of Guam, a single snake weighing down a power line can—and does—cut off electricity to large portions of the island.

By now, the snakes have been, in part, a victim of their own success. The forests, which used to ring with birdsong, have gone silent. The native shrews are almost gone, and even the invasive black rats (of course there are rats) are kept down by the slithering hordes. With less to eat, Siers estimates there are now only between one and two million snakes on the island—about ten for every human on Guam.

Scientists tried everything to find out what might kill them, Siers says. Pesticides on their skin, insecticides the snakes might eat. The surprise winner? Acetaminophen, which you probably think of as Tylenol. (Though, Siers notes, the brand is in no way associated with Guam's snake issue. The scientists get their own acetaminophen and press the pills themselves.)

In high-enough doses, acetaminophen will cause liver problems in basically anything with a liver. But that wasn't what was happening with the brown tree snakes. In turns out, in snakes, acetaminophen stops their red blood cells from taking up oxygen. The net result, Siers explains, is similar to what happens when mammals are poisoned with carbon monoxide. The snake just gets drowsy, goes to sleep, and never wakes up. "It was basically a silver bullet," he says.

Now the snakes had to be convinced to take their medicine. The scientists needed a treat with a truly tempting smell. The first efforts involved tossing dead baby mice stuffed with acetaminophen out of helicopters by the bucketful. But decomposition "rendered [them] unpalatable" by about three to four days after deployment. And of course, they are dead baby mice. Brown tree snakes like them, but humans do not. Tossing dead baby mice out of helicopters had terrible optics.

So Rafael Garcia, an engineer at the USDA Agricultural Research Service in Wyndmoor, Pennsylvania, Bruce Kimball, a chemist at the Monell Chemical Senses Center, and their colleagues set out to build a better bait. It's a bigger challenge than you might think. It needs to hold together and not rot for up to fourteen days. It needs to be meaty and attractive to snakes. It needs to be fatty and oily, and must remain tasty even after it's been glued inside the paper delivery tubes.

As Mary Roach notes in *Fuzz: When Nature Breaks the Law*, what they came up with looks completely indistinguishable from Spam. "The magic in the bait is a mixture of fats that mimic those in the baby mouse skin," Garcia says. They call this magic mix "mouse butter." Delicious.

The baits—dead mouse or Spam look-alike—are laced with 80 milligrams of acetaminophen (about half as much as you'd find in a children's Tylenol) and loaded into paper cartridges and then into small paper tubes. The tubes are wrapped with streamers and stuffed into another tube, and the whole thing is shot out of a helicopter. As it falls, the streamer helps it hook into the branches, where it will hopefully land right in front of a brown tree snake's hungry little nose.

Not that it often does. Even with an automated dispenser shooting streamered snake snacks into the forest, the scientists estimate that the snakes only ever took about 6 percent of the bait. That doesn't sound like a lot. But the test areas were coated with 13,200 baits. So even if only 6 percent get eaten, that's more than seven hundred snakes that got their dose. A follow-up study a month later found that new, nontoxic bait got eaten about 41 percent less than before, suggesting that 40 percent fewer snakes were there to eat it.

It's nowhere near enough, and Siers doesn't think it ever will be. "A lot of times stories come across as '[scientists have been] trying to eradicate snakes from Guam for twenty years and have failed,'" he notes. But the reality is that eradication isn't even an option. "We know how difficult it would actually be."

As it is, the piles of poisoned baits can, at best, be used to keep snake populations down in places where people are most worried about them—around airports or shipping warehouses, which might inadvertently ship snakes elsewhere, areas where scientists are trying to breed the last native birds, and of course, the power stations.

If it could keep the masses of snakes on Guam at bay, I asked, what about the pythons in the Everglades? After all, dead mice out of a helicopter is far more efficient than hundreds of hunters driving the levies at night. Siers looked glum. Acetaminophen will definitely kill every snake at the right dose, he says. But pythons are just too large for this to be a practical measure. The doses used on Guam can fell a juvenile python, but when scientists scaled up their findings, they found that the average adult Burmese python would need to eat more than four grams of acetaminophen. That's a lot of baby mice.

Second, the brown tree snake on Guam is a desperate creature. Desperate enough to consider dead bait. "We're super lucky about what the brown tree snake is," Siers says. "Some brown tree snakes, especially when there's not a lot of alternative prey, will eat carrion." With most native species gone from Guam, the snakes there are hungry enough to take them. But if other prey is around, the ersatz Spam goes uneaten.

Guam had one major thing going for it, Siers notes: there were no other snakes. That means no native snakes might accidentally eat the bait. "Florida's got lots of native snakes that are highly valued," Siers says. Other snakes, with their usual food sources denuded by the Burmese pythons, might end up desperate enough to take the poisoned bait themselves.

The Everglades may have suffered huge declines in rabbits and other mammals. But for pythons, there's still plenty of food, says Kristen Hart, the ecologist at the USGS Wetland and Aquatic Research Center in Fort Lauderdale. "So imagine your python in the Everglades, right, and you have all this prey, and somebody puts a trap out there," she says. "Well, you don't have to go in that trap to get the prey, you've got all the other prey." Fish, alligators, eggs, birds, rats, and of course other snakes are all potential python food. In comparison, some sketchy-looking counterfeit Spam hitting the ground seems pretty unappetizing.

Hart is quick to say she does not do python management herself. She only does the science. But part of that science is figuring out what will tempt a python into a trap. And thus far, the best answer to that is other pythons.

SNAPE HAS BEEN under the same rock pile for ten days. The scientists are out of patience. Armed with crowbars, pry bars, and bare hands, the team from the Crocodile Lake National Wildlife Refuge in the Florida Keys begins wedging up large chunks of concrete, rebar, and coral rock. After a few minutes of sweating, they get to a dirt-covered metal sheet propped over some concrete bricks. Katie King, a conservation associate at the Conservancy of Southwest Florida, hops on top of the pile and scans closely with her telemetry device, hunting for the "bip, bip, bip" of Snape's radio signal.

He's in there all right. With a few more heaves, the scientists lift up the sheet. "There he is!" shouts Jeremy Dixon, the refuge manager at Crocodile Lake. King darts forward and shoves her hands into the dark crevice with cheerful fearlessness.

Burmese pythons have been breeding in the Florida Keys since at least 2016, after riding the waves south from the Everglades. The snakes found plenty to eat, including deer, stray cats, possums, and raccoons. But Mike Cove, previously seen forty miles north with pythons in his pants, is concerned in the Keys with the Key Largo woodrat (*Neotoma floridana smalli*). This rat subspecies, which lives only in the Florida Keys, makes its home in the strips of tropical forest on the islands. It has little in common with the rats of the New York City subway beyond size and a superficial mammalian resemblance. Not only is it native to the United States and a completely different genus, it prefers a vegan diet of fruits, seeds, leaves, and plant buds. It is also a dedicated architect, constructing huge nests out of piles of rocks and sticks. The nests in turn attract other residents, from native cotton mice to invasive iguanas.

But the Key Largo woodrat has had a tough time of it lately, and not just because it's got "rat" in its name. (Cove says that scientists tried calling it the "Key Largo koala" as a PR move. No luck.) The rats had been facing increasing predation from the large colonies of stray cats that roamed North Key Largo. Working together with Ocean Reef's Trap-Neuter-Release program, Cove managed to convince the cat lovers that strays needed to be adopted, and feral cats needed to be contained in a large outdoor enclosure. The Key Largo woodrat populations seemed like they might recover. "In 2017 we were all celebrating and patting ourselves on the back," Cove says. Then Burmese python numbers began slithering upward. Cats under control, Cove is studying ways to get rid of the pythons, and again save the woodrat.

Now, three grinning scientists are wrestling an eleven-foot Burmese python out of his nice safe hole, with King's hands tight just below the snake's head. Snape does not want to go. His midsection writhes back and forth in midair, and his tail wraps around anything—the scientists' arms, legs, or both—trying to get purchase. His pale mouth gapes wide, showing an impressive array of recurved dentistry. Two feet or so behind the teeth, a golf ball–size bulge protrudes from his side. It's an accelerometer, a device that's been tracking how quickly (if at all) Snape has been moving for the past six months. A few feet from the tail, another wire sticks out along his body. This leads to another, smaller bulge—the tracking device that betrayed his location.

A sudden spurt of yellow flies from Snape's vent. The liquid mix of pee and poo hits a rock—as well as two of the scientists grappling with the snake. A pungent smell—like the whiff of ultraconcentrated urine you get if you accidentally lean over a porta-potty hole at the wrong moment—fills the air. This musk is most snakes' last (best?) defense mechanism. This Burmese python is almost three meters of abs, equipped with needle-sharp teeth at one end. But when all else fails? Poop yourself.

The scientists are well acquainted with this tactic. Dixon runs for a large white sack, and two handlers form Snape into a large U. The bottom of the U goes in first, then the tail, and finally the head. With an expert twist, Dixon ties off the bag. The last scout snake is in for the summer.

Snape is one of several snakes equipped with tracking devices and accelerometers, released in South Florida in the fall of 2020. Their mission? To find more snakes.

The latest effort includes Snape and his companions, an expansion of a program started by Hart on the Florida mainland. When Hart was looking for a Burmese python's weakness, she initially came across a problem: they don't really have any. "They're totally adapting to the environment here," she says. Even salt water—usually a death sentence for most snakes, and what Cove and others were depending on to keep the Florida Keys python-free—can't stop them. In fact, Hart and her colleagues have found that Burmese python hatchlings can survive for months in brackish water, and even get through a month in salt water. That's on average—one hatchling was cheerfully still feeding, in salt water, after two hundred days.

Then of course there's the problem of catching them. Kalil and her teammates hunt on the roads and levies. But the Florida Everglades doesn't have a lot of roads. "Often we're helicoptering into places like it's the Wild West," Hart says. "Except it's not, it's here in Florida and coated with mosquitoes."

When one finally does catch a python, there's the issue of killing it. "They also have this weird ability to regrow their organs," Hart explains. You can't design something that will attack the heart or liver. They'll just grow another. "How can you get around the ability of an animal to regrow organs?" Right now, Kalil and other hunters rely on fully destroying the brain—and sometimes decapitation for good measure. Thus far, pythons are not regrowing heads.

I ask if the pythons really need to die. The Burmese python is actually listed by the International Union for Conservation of Nature

(IUCN) Red List of Threatened Species as "vulnerable" in its native range, which includes places like Bangladesh, Cambodia, Vietnam, Thailand, and Indonesia—Southeast Asia, basically. There, the snake is hunted for food, medicine, and so people can make fashionable boots and bags out of its skin. It's also threatened by habitat destruction. If we've got too many pythons here, why not just catch a bunch and send them there to help with conservation efforts?

Hart explains that the snakes are already adapting to life in Florida quicker than a retiree from New Jersey. And the Floridian Burmese pythons might never have been the same as the Southeast Asian populations to begin with. "Presumably, they didn't start out the same because they were part of a pet trade," she says.

She's already noticed differences. "I think their reproductive cycle is probably way quicker here," she theorizes. "They potentially breed every year, whereas in the native range, it might be every third year." Her lab has also shown that the Florida Burmese pythons (*P. bivittatus*) mated at some point with Indian pythons (*P. molurus*). As to whether the Everglades pythons are changing enough over time to have substantially different DNA, she says, "Ask me five years from now and we might see an answer."

Hard to hunt, tough to poison, tough to kill. Hart was left attacking the only weakness she could find. This is the goal of the scout snake program, which is an adaptation of something called the "Judas" technique—the idea that one tagged, clueless animal will seek out others and lead scientists right to them (they call it "scout snakes," Hart notes, because the word "Judas" was apparently offensive). The only hitch is that pythons are not social creatures—unless they are looking for love.

During the mating season, the snakes don't eat much. All their effort is focused on finding other snakes. A female releases a sexy scent, and males come from far and wide. The season culminates in an ecstatic reptilian orgy, with one female and several males. "There's like a mating ball," Hart says. "So it's difficult to go in and break that up, but it has occurred." I have a lovely vision of a bunch of scientists hiking into the brush and scooping up a big ball of very distracted pythons with a backhoe. Hart says it's never like that. The snakes scatter immediately, and "a lot of the time this is occurring on the edge of a canal, so you're getting what you can grab as quick as you can," before the pythons slip into the water and out of reach.

The tracking devices are expensive, and the process is slow, but sometimes more effective than hunting with floodlights night after night, because it increases the odds of finding and removing a female that might make more pythons. And so that's what the team at Crocodile Lake are trying. Armed with a tracking device, the scientists release horny snakes during the mating season. (All the Key Largo snakes have names from the *Harry Potter* Slytherin house— Snape, Draco, Crabbe, Bellatrix.) When there is enough data to determine where the snakes are hanging out during the breeding season, the scientists set off in pursuit. With any luck, they'll find a scout snake in the midst of a multisnake romantic encounter. At the end of the mating season, the scientists track down any remaining tagged snakes who did not get laid and take them back into captivity before they get too hungry and cause more woodrat fatalities.

In some areas of Florida, this method has had some success. In Key Largo, not so much. Snape, for example, was captured all alone. It could be a good sign, Cove says. Maybe the Slytherin scouts can't find other snakes because there aren't enough other Burmese pythons to make up a dating scene. Or maybe the snakes they released were just unlucky. It will take more time—and probably more scout snakes—to find out.

TENS OF THOUSANDS of Burmese pythons in the Everglades, and we are relying on hunters with floodlights and snake matchmakers to try to catch them. Hart says maybe genetics will have the answer, in the form of gene drives. These are genetic alterations that could help render an animal infertile. Some have already been tested in mosquitoes—gene drives that ensure only nonbiting males get born, with the goal of driving the insect to local extirpation. But that's in mosquitoes. It's not even something scientists can do reliably in mice (which we will get to). Pythons are going to take a while.

Kalil holds out hope for other technologies. Artificial intelligence, with its skills in pattern recognition, might be able to gather enough data to reliably tell snake from stick. Drones equipped with near-infrared cameras could fly where humans can't drive. "Maybe they'd have grenades," she jokes. "Boom."

The Florida Python Challenge is now an annual event. It will bring more people to South Florida to hunt, but Kalil worries it could be a mixed blessing. She has many stories of people who came out to hunt, trying to conquer their fears with herpetological slaughter. But they might not be able to tell nonnative pythons from the many native snakes that also slide past their floodlights. Novice hunters without proper training could also end up torturing the snakes instead of humanely killing them.

I'm worried about the incentives too. Something called the Cobra Effect comes to mind. When the British government controlled India in the nineteenth and twentieth centuries, they were concerned about the large numbers of highly poisonous cobras in Delhi. They offered a bounty for the snakes, hoping Indians would cull their numbers.

Remember the ill-fated rat bounty in Vietnam? People in India reacted the same way. They killed wild cobras, of course. But they also soon began to breed them. More cobras, more money. When the British government found this out, they stopped the bounty—and the cobra breeders released their now worthless charges into the countryside.

Burmese pythons won't ever go back where they came from.

But the way people deal with them—the instant fear and quickness to label them a pest—is part of how we grant ourselves license to kill.

The very word "pest" confers a type of "epistemic violence," says Francisco Santiago-Ávila, a conservationist at Project Coyote and the Rewilding Institute. It gives people the idea an animal is out of place, unworthy—or at least less worthy than the other things in the area you want to save from it. The word helps us take away any tugs that animal might have had on our conscience. A pest isn't part of nature anymore. It isn't wildlife. A pest becomes an evil influence that must be eliminated, and the ends will now almost always justify the means.

Kalil has always liked snakes. She respects every python she catches. But she still thinks they have to go. "Basically, every single one I take out of the environment is saving hundreds of thousands of lives. That snake is out here and it doesn't belong," she says. "[That's] the way I'm able to sleep at night." She's already seen the changes pythons have wrought over her time in Florida. "Forty years ago," she says, "we would have seen probably about a dozen animals dead on the road already."

At first this seems like a non sequitur. But it's not. If there aren't a lot of mammals living in the Everglades, there won't be a lot of mammals dying on the roads. And it's true that as we drive I don't see a single roadkill. In the dusk, we see a small herd of feral pigs, and the darkness reveals a few rats. But no other mammals disturb the swamp. It's a chorus of frogs, birds, alligators, insects—and snakes.

Driving along with Kalil in her python skin–lined truck with the cool, humid breeze, the Everglades begin to seem so impossibly vast, extending out forever into the darkness. In all that night, more than eight hours of searching, we caught one snake. That's pretty typical, Kalil says. I ask whether, in the face of thousands of pythons she will never catch, she will ever stop trying.

"Never," she says quietly, after a long pause. "I'm never going to stop trying to kill every Burmese python in the Everglades. That would be giving up hope." Kalil and her Avengers will keep going out in the night. They will drive the levies and hold on to the Everglades that she remembers—and keep working for the one she hopes will be. One snake at a time.

Burmese pythons have already changed the Everglades. They have changed it forever, even if the python hunters got rid of them tomorrow. Odds are they never will. But the ways people do it, the approaches they use, are bound up with Western culture's understanding of snakes as evil and scary, and the belief that the spaces humans value are gardens of Eden, places where evil should not dare to tread. And when animals like snakes invade, conservationists and hunters alike won't hesitate to throw their cultural weight toward destruction—to do almost anything to save paradise.

PART II

A PLACE TO CALL HOME

3

A NEST OF MICE

I spent my formative years in a lab studying mice. I was researching the effects of drugs on the brain—from stimulants like cocaine to antidepressants and hallucinogens. I was trying, in the small, creeping millimeters of scientific publications, to work out the pathways in our heads, the electrical firings and misfirings, that give us pleasure and pain, ecstasy and misery.

It takes only days in biomedical research to learn to have your key card in easy reach at all times. Attached to a little extender on my pocket or belt, or hooked to a lanyard, my dorky little card was sure to add a jarring note to every outfit. But I needed my nerd cred, first to get in the front door and past the security guard and then to get into the hallway leading to the offices. Then again into a bright white room humming with equipment. Some of the whiteness came from the floor, some from the walls, and some from the clean bench paper carefully taped to every horizontal surface.

Several times a day, I'd head past loudly humming freezers and put on a blue disposable lab coat. I'd wriggle my hands into nitrile gloves with a practiced twist. (Snapping gloves is for amateurs. Slide, then twist.) I'd add a pair of protective booties over always, always closed-toe shoes.

Another key card entry, another pair of booties slipped over the first pair. A hairnet, a surgical mask. A final key card entry into a long gray hallway with the constant tang of 70 percent ethanol cleaning fluid. It was easy to miss the slightly downward slope as I walked through the building. Once in the windowless hallway, though, the subterranean feeling became obvious. Under fluorescent lights, past rolling carts, and through a heavy metal door. The room was always dim, filled with steel racks and plastic cages, with a constant dry, rustling, scuttling sound in the background. I'd breathe in the dusty, earthy smell of corncob, wheat, and a little bit of pee.

It smelled like home.

I love mice. The feel of their tiny little claws as they climb over my hand. The soft fur and tiny tummies. The bright eyes and whiskers that flare forward enthusiastically when they get treats. Once, two colleagues caught me working with my mice on the weekend—and singing to them. I especially loved to give them Froot Loops. Watching a mouse eat a Froot Loop was like watching a human try to eat a car tire. They'd roll it around, nibbling away at the edges, finally breaking through and devouring the center. Then they'd flop out in the cage to sleep it off. No regrets.

They were mice, sure. But a lab mouse in a cage—beyond the same rough size, soft fur, and shining eyes—bears little resemblance to the mice that plagued the cabin of my friend Eva. Eva is a journalist who had brought her family over from Germany for a yearlong fellowship at MIT in 2019. In March 2020, she found herself trapped in a two-bedroom apartment in Cambridge with her husband and three kids, who could no longer go to school due to the COVID-19 pandemic. Desperate to find a little breathing room, Eva and Stefan crammed the kids into a rental car and drove to a cabin in western Virginia.

The mice were there to greet them, and only too happy to savor Eva's professional-level sourdough bread. Soon, she was texting me pictures of nibbled loaves and the mini poops of mice on the cabin's kitchen table. Worried about hantavirus, the family put all their food in the oven. It didn't work. The mice slid right in. The only safe spots turned out to be the fridge and microwave, stuffed every night with as much food as they could hold.

Both of these places—a cabin in the woods and a sterile laboratory—offer a niche that mice have filled. One is as old as our first attempts at civilization. Humans have had house mice since we had houses, and we've been leaping on chairs to get away from them probably ever since we had chairs to leap on. In the pest niche, mice make a living off our leavings—a living so successful they've spread across the world.

But the other niche—the lab niche—is relatively new. Here, the mouse isn't a pest. Instead of living off us, we are living longer and healthier lives off them. We thrive from the data mice produce and the lessons they can teach us. As living laboratory tools, mice don't just advance our knowledge; they change science itself—molding what questions we ask and how we look for the answers.

And yet, that new niche would never have been possible without the old. First, the mouse had to live among us. It had to become a constant in the human landscape. It became both so common as to seem worthless and invasive enough to be a constant minor irritation.

A mouse isn't a human. But it lives in the human world and eats human food. It navigates life with a mammalian brain, with a physiology very like our own. It's something we recognize as clever, and even cute—but which many people would kill without a second thought. And so, when we went to look for something to stand in for us as we learned the secrets of our bodies, something living but not too lovable, the mouse was a natural choice. Just human enough to be not human at all.

SOMETIME BETWEEN FIFTEEN thousand and eleven thousand years ago, there was an odd group of humans. Just a dozen or so families, probably with their hunting dogs. This group of humans was into something a little kinky. They were into long-term housing.

The first settlements were not towns or even villages. Just a few roughly circular huts in the area that is now Israel, Palestine, Jordan, and many other countries. Often, when people think of the Middle East now, we think of something desertlike. Drier scrubland, some oases. But fifteen thousand years ago, it was wetter than it is today. There were lots of long grasses that people could eat (including the plants we would domesticate as our wheat, barley, and millet). The hills had forests of oak and pistachios. The valleys in Jordan had marshes. The grass was full of gazelles that humans might even have tried taming at one time.

This is where humans first tried housekeeping. Their stone huts weren't occupied all year round. Instead, they were probably seasonal stops—more of a summer home or winter home than a temporary camp. The people living in these huts weren't farming yet, though they were baking the first bread and brewing the first beer (priorities!). They still got their grain by hunting and gathering, switching locations to follow their food.

But they had a mouse problem. Or rather, one mouse species had a human solution. Lior Weissbrod has studied the remnants of these mice, not entirely by choice. His adviser, Mina Weinstein-Evron, was working on prehistoric sites in the Galilee when Weissbrod arrived to work on his master's thesis. By the time he was ready for research, all the large animal bones at the site had been claimed for study. Mice were "the only material that was still waiting for someone to work on."

An archaeologist at the University of Haifa in Israel at the time, Weissbrod was interested in the early relationships between people and animals. He ended up finding an unexpected one in the tiny teeth mice left behind. It turns out that when humans first moved to the area and built their stone huts around fifteen thousand years ago, two species of mice were already there, competing for food.

Life at a mousy scale has never been easy. Every grain was worth a scuffle, and every hole could be a home—or a place where a predator lurked. Maybe one species, the one we now call *Mus macedonicus*, was a little warier than its closely related neighbor. For this mouse, paranoia made sense. Everything was out to get you. With more fear, *M. macedonicus* might miss out on some tasty, risky treats, but it might also stay safe from lurking hawks or cats. Another species, though, was after the same food, in the same places, resulting in constant competition. A low-level mouse war continued for generations.

Then humans arrived, and *Mus macedonicus* fled. It feared loud human voices and big human feet. And when one species flitted away, its rivals saw a chance—and pounced.

This new mouse—which we would come to call *Mus musculus domesticus*—didn't seem to mind people so much. Maybe it had always been a little bolder and a little less cautious. It got those tasty treats that *M. macedonicus* left behind, but it was also probably a more frequent snack for the local predators.

To this risk-taking mouse, humans brought noise, but they also brought opportunity. Their stone huts offered shelter in the spaces and crevices—shelter predators couldn't access. Human diets left piles of trash nearby. That trash had plenty of food for a mouse brave enough to take advantage.

The human settlement was only a few houses, but when you're the size of a mouse, a few paltry huts are a vast landscape. "Once you get humans into the picture and human settlements . . . there's a creation of a new habitat," Weissbrod explains.

The new mouse slipped into the cracks of our new world. Until the humans left. The settlement ended up abandoned for about a thousand years, and the murine struggle resumed. During that time, M. macedonicus came out of hiding, replacing M. m. domesticus. M. macedonicus's skills were much more well matched to the returning natural landscape.

But humans always come back. When they did, they brought back their noise, their trash, and their opportunities. M. macedonicus turned tail. M. m. domesticus came out on top once again.

Tens of thousands of years later, M. macedonicus is still a wild mouse, living between Eastern Europe and Israel. Its wildly successful cousin, on the other hand, occupies every continent except Antarctica. Its bold behaviors were a speciation event, separating it from a common ancestor that never happened on the human opportunity. Mus musculus domesticus had won the evolutionary lottery to live with people. The house mouse was born.

What was it that M. m. domesticus had that M. macedonicus didn't? "That's the big question," says Thomas Cucchi. Cucchi, a zooarchaeologist at the National Museum of Natural History in Paris, specializes in ancient animal remains, especially those associated with some of the first human settlements. He supervised Weissbrod's final dissertation studying ancient mouse teeth.

The house mouse's speed, intelligence, and genuine moxie has made it the most successful pest on the planet. Yes, even more than rats, Cucchi says. "They are found everywhere, everywhere." Arctic island or subantarctic, it doesn't matter. "They're less noticeable, but [like rats] they're also very successful at adapting to new environments." After all, if they're not running directly across your line of sight, it's easier to leave them be than a rat, he says. "As long as the human presence is there, you'll find a mouse."

The real secret here lies in evolution, of course. When a male mouse and female mouse love each other very much . . . nah, never mind. When a female mouse can tolerate a male mouse for the few seconds he needs to do the deed, twenty-one days later, they will produce a litter of pups—anywhere from four to twelve. In another twenty-one days, the pups have gone from looking like small wads of chewed gum with legs to hyperactive teen mice ready to be weaned and given learner's permits. Within that time, the female mouse can get pregnant again, giving birth to the next litter when the first is weaned. In another three weeks, that first litter will be ready to go off and make grandmice. If a single pair of mice produce twelve babies in a litter, and ten litters in a year, one pair alone can produce 120 mice. While all that is happening, their older kids are going off and having more kids, and those kids are also having kids. It adds up.

Each generation of mice would have more genes that produced traits that made it slightly easier to survive around humans and fewer that made their lives harder. With statistics like that, it would be reasonable for those genes to spread within a population in a few thousand years, much like the recent human ability to digest lactose as adults. In terms of evolution, that's an eyeblink.

But to Tom Gilbert, it's not nearly fast enough.

Gilbert is an evolutionary biologist and director of the Center for Evolutionary Hologenomics at the University of Copenhagen. For mice to quickly take advantage of a human environment, he says, it's not the mouse that needs to change. It's the mouse's microbes.

By now, most of us know we aren't alone in our journey through life. Forget the mice in our walls or the spider in the corner. We're walking ecosystems, from mites plugged facedown in our pores, eating our skin grease (and then pooping it out. On our faces. Delicious), to the trillions of bacteria in our guts. And if a single pair of mice can produce a family of 120 in a year, that's nothing to the numbers of bacteria that can be pumped out in the same time span.

But what determines which bacteria are teeming in our bodies? Gilbert believes that, in part, our genes do. By controlling the development of our insides, our own genetics can create a more or less welcoming environment for microbes. In each person's gut, some microbes might be at home, and some might be less so. And this varies not just from species to species (mice and chickens probably don't have the same gut microbes), but from individual to individual. Each person's (or animal's) tiny genetic tweaks might translate, Gilbert believes, to a personalized microbiome that can affect the larger organism's behavior.

It's not just the genes that talk to the microbes. The microbes talk back. Some of their chemical messages are easily read by other microbes, as they form alliances and wage minor wars in your gut. Others are messages for the microbes' host. Some messages are simple enough, influencing our bodies to make the gut a cushier place for that particular bacterium. Other messages can influence far more than local cells; they can alter the behavior of the whole animal. For example, treating mice with lactic-acid-producing bacteria in their guts can reduce the amount of stress chemicals mice produce in their blood. They also showed less anxiety and depression-like behaviors—they hid less in new, scary environments. (Unfortunately, attempts to produce the same effect with the same bacterial strain in humans haven't worked yet.) From tweaking the chemical signals in the nervous system, it's a small step to changing an animal's behavior, courtesy of the pharmacy in your gut. "All microbes are little drug factories," Gilbert says.

Those microbes could have been the first step in putting the "house" in the house mouse. "The microbiome does play a role in mouse behavior, like it plays a role in our behavior," says Jane Foster, a neuroscientist at McMaster University who has spent her career transferring microbes from mouse to mouse.

Foster notes that microbes can get transferred both vertically and horizontally. Vertically, meaning that mothers pass on microbes to their babies. Horizontally, meaning that we (and mice) pass them to each other. In the case of humans, that means by way of the little bits of fecal bacteria left on our disgusting hands, as we touch doorknobs, kitchen counters, and each other.

In mice, they cut out the middleman and eat each other's poo. It's called coprophagy, and it's not nearly as unpopular as you might think. Lots of species, from dogs to horses, do it. Even gorillas indulge in a little fecal feeding.

It's a great way to share gut bugs, and those gut bugs can determine behavior. For example, mice with no gut bacteria at all are just more chill. They are unafraid to hang out in the open arms of an elevated maze (normally a scary proposition for a prey animal that likes darkness). Other studies have shown that different lab mouse strains react differently to a bacterial dose. Some show effects that look like they've been given antidepressants, while others have no response at all.

What if those ancient mice—*Mus macedonicus* and *Mus musculus* (soon to be) *domesticus*—also responded to gut bugs in different ways? What if, for example, they both took a few bites of human poo. The bacteria would flood their systems. In *M. macedonicus*, they might never quite manage to settle in. The bacteria wouldn't be very comfortable in the *macedonicus* gut. The microbes might have pumped out chemicals, trying to communicate with their hosts, but *macedonicus* didn't listen. It didn't relax.

But in *domesticus*, maybe, the bacteria found a better home. The *domesticus* gut was a friendly one, warm and just acidic enough, with other friendly bacteria who made good neighbors. The new bacteria began pumping out signals. Those chemicals said, "Fear not. Stay here. This is a good gig if you can hack it."

Domesticus stayed. It pooped. Other *domesticus* ate the poop, or more human poop, and got their own dose of the feel-good bacteria. This is all a guess, and it's possible that we'll never know what really took place. But it is an intriguing idea that a few good bacteria could make evolution fly faster than genetics alone could ever account for.

THE MICE AND their microbes weren't just evolving the ability to rest and relax in a human environment. They were also developing new talents.

It takes skill for animals to thrive in our presence. For one, it helps not to be too picky, says Anja Guenther, a behavioral ecologist at the Max Planck Institute for Evolutionary Biology in Germany. Guenther studies mice. Not the tame, sweet, and somewhat dim lab mice I used to let crawl around in my lap. She breeds and studies wild mice, taken from bucolic rural areas or a rough life in the parks and streets of Berlin. "Most people ask me how [my research] can be related to humans," she says. "My usual answer is, I'm not interested in humans."

Guenther wants to know how animals like mice adapt in order to live in human environments. She wants to know what it takes to live the life of a pest.

Mice, along with many other animals that love human environments, such as seagulls, pigeons, sparrows, rats, and raccoons, are generalists. In the wild, mice eat grains, insects, fruits, and nuts. But does it really matter if the grain is on the stalk or in a leftover croissant? If a mouse isn't too demanding, the human world has a lot to offer. "We nicely offer animals year-round stable food resources, if they can find a way to get to these food resources," Guenther says. "We offer nesting places in our houses . . . in everything we construct, basically, there are always niches for the animals where they can nest."

But to make it, a mouse needs to be more than a generalist. It needs to be flexible. The human environment is constantly changing. New buildings and new people with new schedules. Just when a mouse finds a box of cereal, the human invests in Tupperware.

It behooves a mouse to develop behavioral flexibility—the ability to adapt to this constantly changing world. And they do. Guenther has shown that humans—or rather, our cities—are making mice streetwise.

Guenther and her colleagues collected three subspecies of mice, M. musculus castaneus, M. musculus musculus, and M. musculus domesticus. All of them had been living with people for between four thousand (castaneus) and thirteen thousand years (domesticus). The scientists offered them a series of problem-solving tasks, each with a tempting mealworm as a reward. To get the prize, the mice had to pull little wads of paper out of little plastic tubes, pull the lid off a petri dish, and even open the window of a tiny LEGO house. The longer the mice had lived with humans, the better they were at solving the tasks. M. m. domesticus—our longest-standing housemate—beat the other two subspecies paws down.

The human world is one heck of an education. When Guenther conducted a similar study on a different genus of mice (Apodemus agrarius), this time comparing city and country mice of the same subspecies, she found that city mice were better at the problem-solving tasks than the poor country mice.

These problem-solving skills are what Guenther describes as "cognitive enhancement." It's not really about mice being smarter. They're not taking anyone's slot on Jeopardy!. They're just always getting better at living with us.

MICE AREN'T THE only species that has brought problem solving to a new level. Gray squirrels, yes, my dear F**king Kevin, solve difficult puzzles more quickly on the first visit when they're raised in urban environments, possibly because they're more willing to try new things. But perhaps the most well-known example is the Eurasian blue tit (Cyanistes caeruleus), a small, chubby bird with a yellow breast, blue cap and back, and a distinctively sassy facial expression. These birds successfully figured out how to either peck through the foil tops or flip off the wax board covers of milk bottles. They were first documented doing this in 1921 in Southampton, England, and the behavior spread all over England, Scotland, Ireland, and Wales. Other tits did it too, as long as old-fashioned dairy deliveries existed.

But really? Milk tops? Petri dish lids? What is this, amateur hour? For real genius, look no further than the bandit face masks and clever paws of our favorite trash panda, the raccoon (*Procyon lotor*), particularly the famous raccoons of Toronto.

For a start, no one knows how many raccoons there are in Toronto. Residents only know there are more than they would like. "It's pretty much impossible to tell," says Suzanne MacDonald, an animal behaviorist at York University. "I would need millions of dollars to be able to do an actual density survey in Toronto."

At some point MacDonald hazarded a guess that there might be a hundred thousand raccoons in the city. But in reality, she has no idea. "I'm usually wrong about these things," she says, "because raccoons constantly surprise us with how much more social they are willing to be in urban areas than we had thought."

One reason people think there are so many raccoons is because their effects are so apparent. In fact, they are so good at getting into people's trash cans (and pet food, and attics, and kitchens) that the city devoted thirty-one million (Canadian) dollars to design and deploy raccoon-proof bins for the city.

"I volunteered to test the prototypes that they were considering," MacDonald says. She tried them on the hardened trash bandits in her own backyard. All three cans worked, but two were not just hard for raccoons, they were too hard for humans. Make a trash bin with too many steps to lock it, and people give up before wildlife does.

The winner was a bin that, in theory, required opposable thumbs to open. There's a twisting motion required to lock the top.

When the garbage truck comes, it flips the can upside down and the weight of gravity opens the bin.

Toronto residents looked forward eagerly to a raccoon-free future, but a few days later, the raccoons were back on top. The designers planned for the raccoons to try to "solve" the problem of the bins. But following the rules is for suckers. All the raccoons had to do was tip the bin over, breaking the lock in the crash. The delicious, delicious garbage was theirs for the taking. "There was much hullabaloo, with the journalists in Toronto going 'Oh, the raccoons figured out the bins!'" MacDonald says. "Well, they didn't figure them out. They just broke them."

These urban raccoons are adapting. In a small comparison of orphaned baby raccoons, MacDonald has observed that those from the city—like city mice and country mice—outperform their rural relatives on problem-solving tasks. Urban raccoons are quicker to try burgling a trash can for a food reward.

"The city is actually creating this überraccoon," she says, just as with mice or blue tits or squirrels.

RACCOONS, EVEN IN Toronto, have plenty of fans. Mice do too. As I watched videos of Guenther's mice nose their way into LEGO houses and pull lids off tiny petri dishes to get a snack, it's really hard not to root for them. When I was a scientist, I knew my mice would die. In fact, I knew I would be the one to kill them. I needed their brains, their blood, their data, and yet I couldn't stop caring. I rooted for them to solve the mazes, be brave and get their tiny rewards. Even when I ended their lives, I mentally apologized, a small "I'm sorry, buddy" to each and every one.

Even when they get in our houses, mice have things to teach us, Harry Walters reminds me. Walters is an anthropologist and a Diné elder, from the Navajo Nation. The Diné have lived around mice and rats for as long as anyone else. Their rats are not *Rattus*, but *Neotoma*, the pack rats native across North America. And their mice aren't *Mus*, but *Peromyscus*.

These rodents aren't fully commensal, like the house mouse is. They're perfectly happy to live on their own in the desert. On the other hand, they won't turn down your woodpile or kitchen. They've got that same flexibility that *Mus* and *Rattus* have and know a good thing when they see one.

If they do make an appearance, Walters says, what is different is who is at fault. Many people might see a mouse in the kitchen and feel outrage. How dare this animal invade my space. How dare it show itself. Get out the snap traps.

Diné do the same, Walters says, but the understanding of what is happening is different. "They don't do it to annoy us, they do it to survive," he says. "They're just trying to live. Doing what is natural." In Diné understanding, he says, animals and humans have a mutual responsibility to respect each other. "In nature they steal nuts, berries, roots, and eat insects," Walters says. "But if we leave trash and garbage they come around. The thing to do is keep your house clean."

A mouse in your kitchen is carrying a message. It's violating its part of the bargain to stay in the wild. But with its arrival, it reminds you of your end of the bargain too. It will do its duty and stay out, but it's your duty to keep your house clean and make it a place a mouse wouldn't want to go. The mouse is not inherently lesser. It has talents—talents for getting into your stuff. Humans in turn must exercise their talent—their rationality—to keep the mice out.

Just because an animal has to be kept out doesn't mean that an animal is worthy of hatred or disgust. Relationships between people and animals have always been far more complex than that.

Mice or rats, for example, might be there to munch on your grain, but isn't that a compliment? It means you have grain to be munched on in the first place. Even ancient peoples, Cucchi says, might have seen the softer side of mice. "Maybe people weren't looking at those mice as pests," he notes. "In places like India, mice and rats around aren't bad, they're lucky," he says. "You have mice and rats around, you have food."

As an example, Cucchi shows me a small stone, about three centimeters long, maybe half the length of an AA battery. The white rock is beautifully and deliberately carved. On one side, there are round, erect ears. In front of the ears, a divot for the eye. On the other end, a delicately pointed nose. There's even a line drawn for the mouth.

It's instantly recognizable. It's a carved head of a mouse. In the back, there's a hole, and it's clearly meant to be worn as a pendant or charm. It looks like something you could buy today on Etsy. But it's from Syria, and it's eight thousand years old.

By that time, humans were farming and herding full time in the area, and M. m. domesticus was a constant companion. And maybe, as this pendant shows, not always an unwelcome one. I can imagine an early Neolithic farmer, those thousands of years ago, home from a hard day's work. In the firelight, they turn a small white stone in calloused hands, carving this tiny mouse, a gift for a beloved child.

That's just my romantic imagination, of course, but Cucchi notes that people don't put time and attention into carving representations of things they have no respect for. The mice might be pests in their grain stores, he says, but making art of mice shows those people saw something else too—something small, clever, and cute.

Even today, there are mice we smile about. Mickey Mouse is the head of a media empire. People wear mouse-ear hats. There's an entire children's book series, Redwall, that captivated my brother and me as children. In it, heroic mice live together as monks in an abbey, fighting off vicious rats, stoats, and other nefarious characters. They're like rats, but cuter. Softer. Smaller. In a black-and-white moral world, mice come in all shades of gray.

When someone carved that tiny mouse-head pendant those thousands of years ago, the animal was already in our imaginations. Even then, people had that little tug, that tiny heartstring pull that said, "This creature is alive, just like you."

THIS HAS NEVER stopped us from trapping and killing mice, of course. We created a niche that they evolved to fill, and mice started changing us immediately in return. They took up our time and our thoughts. We built granaries to keep them out, invented Tupperware and cans, and, of course, mousetraps.

The oldest trap we know of caught mice nearly forty-five hundred years ago in the Indus Valley civilizations in what is now Pakistan. Cucchi has another version, shaped like a classic pottery jar, tipped on its side. It also has a long pottery handle at the top, making it easy to lift. Both the ancient trap and the modern version have holes punched deliberately in the sides to let the water in when you dunk your trapped victim in a bucket.

These traps don't bear much resemblance to the snap traps most of us think of when we hear "mousetrap." If you want to see those, head to Galloway, Ohio, where Tom Parr, the president of the North American Trap Collectors Association, runs the Trap History Museum. The big building used to be Parr's medical supply business. Now, the entire bottom floor is filled with traps for everything from coyotes to bears.

The mice have a room to themselves, where cage traps, snap traps, live traps, multicatch traps, and poison baits are neatly stacked, layered, and hung on every available surface.

The first American mousetraps weren't pretty pottery affairs, Parr explains. "They weren't a contraption, it was like a rock stood up on its side and a trigger mechanism, like three sticks, and when the animal pulled on a stick it would collapse." That box held up with a stick that many people think of when they think of "trap."

The famous wooden snap trap didn't come along until 1894, the invention of William Hooker of Abingdon, Illinois. He and others quickly found there was money to be made in mousetrapping. "He patented one of the first mousetraps, that was copied over and over and over with just simple little variations that made it still patentable," Parr says. The same went for cage traps or even traps made out of the now trendy mason jar, each with a slightly different mouse catching lid.

As traps proliferated, books about them did too. Most are by the late, great trap collector David Drummond, a man recognized as a pillar of the mousetrap collecting community. He is also the author of books such as *Nineteenth Century Mouse Traps Patented in the U.S.A.: An Illustrated Guide.*

It's the names of the traps that make the reading really fascinating, and it's also the names that make snap traps Parr's personal favorite. "There's just a whole bunch of names," he says. "Knock 'Em Stiff. Tomcat." A perusal of Drummond's works quickly brings up others: Buffalo Bill. The Last Word.

Each has a mix of humor, desperation, and a certain amount of bloody-mindedness. There's no disguising that these traps are designed to kill. But there's also the acknowledgment that if you buy these traps, you have a scurrying problem that you are desperate to solve. Most catch one mouse at a time. A few will lure in multiples. This seems like it would be plenty for a kitchen or, spread out, for a barn.

It's not nearly enough for Australia.

Even though mice can multiply at lightning speed, we don't really see it happen often in North America. Predation and infant mortality keep the numbers down. But when Australia achieves the right conditions, the mouse population explodes. The condition is called a mouse plague, and it is a problem that has, er, plagued Australia since they were first reported, about one hundred years after white colonists brought the mouse to shore, probably with the First

Fleet of British colonists in 1788. (Plagues before that were of rats, not mice. Australia has several native species of mice and rats, one of which, *Rattus villosissimus*, has been known to exist in plague-like proportions itself. Australia is a very special place.)

Mice quickly spread across the continent, aided by Europeans' movement and propensity to farm delicious grains. And then, the climate of Australia came into play: It's not just heat or cold that determines whether there will be a mouse plague. It's rain and grain.

"In Europe it rains frogs, and Frenchmen are thankful for them; here at the antipodes we have mice in showers," reads an excerpt from the Pastoral Times in a newspaper called the *Queanbeyan Age* in 1871. Nestled amid reports of fancy new clocks for the post office, "lady students" for the university, and a child who fell through the seat of an outhouse and nearly drowned are reports of mouse plague. "Nothing comes amiss to them, and what we are to do I know not," the excerpt read. "I and my brother have a trap by which we have captured twelve hundred and three at a time in an evening."

Australia is famous for its extensive droughts, but for there to be a drought, there must also be rainy times to compare it to. In New South Wales, for example, the months between January and June are the wettest. If there's enough rain, farmers will get bumper crops of wheat, barley, and canola (also lentils, but apparently mice don't like lentils).

As they collect the harvest, farmers will miss grains. "One of the things I've been asking the farmers is, '[Do] you know how much grain is going out the back of your header?'" says Steve Henry, an ecologist at the Commonwealth Scientific and Industrial Research Organisation (CSIRO) in Australia. A grain header is a machine for harvesting grain that sits in front of the combine, cutting the grain and feeding it into the combine for collection. In the process, some grain always gets dumped on the ground—a buffet for mice. Sometimes, Henry says, whole tons get left behind. "And you'd be surprised how many farmers don't measure it." Even when farmers are fastidious about collecting all the grain, large wind events can scatter tons of barley off the stalk. This is bounty that helps sustain mice, living quietly underneath the soil. If there's a decent crop in the fall, with missed grains coating the ground, and a nice autumn rain, and animals survive through the winter, when spring is come, mice are sprung.

"The ground's moving with mice when they're really bad," Henry says. In a 1917 outbreak in Victoria, an estimated thirty-two million mice were caught.

Videos of a recent outbreak in 2021 show hundreds of tiny figures, scurrying and leaping through the darkness. There's a constant rustling. The pitter-patter of thousands of very tiny feet. "In a really monumental outbreak, they talk about a carpet of mice on the roads," Henry says. A carpet of dead mice, the ones that didn't look both ways. There are reports of mice biting patients as they lay in their hospital beds.

But the truly awful part is the smell. "It's basically the smell of rodents in general," Henry says. Mice mark their territory and communicate through scents in their pee. But eau de mouse plague has a darker underscent. Henry says he gave a talk at a local sports club in Walgett, for instance, a town of around two thousand people in New South Wales. The town had a bad case of mouse plague, and "the smell of death when you walk into the club was unbelievable."

If mice are born, mice must die. At first, the Australian plague mice die from traps, as people deploy buckets, snap traps, and more to protect their houses from the horde. Then, they die from poison. The farmers put out bait coated with zinc phosphide, which turns to toxic phosphine gas in the mouse gut. Near their houses they will use advanced types of anticoagulants such as difethialone, which kills mice just as well as it does rats.

The animals take the bait and head back to their burrows. They die underground—often under or inside the walls of buildings.

There, the hundreds of furry bodies rot, and the farm, house, and town fill with the stench of death.

Poison or traps aren't what end a mouse plague. No one knows what does. Some mouse plagues last weeks, some months, until winter gets cold enough to, er, chill out the population. Just as mysteriously as the millions appear, it seems, they are gone. "I get phone calls from farmers saying, 'Where have all my mice gone?'" Henry says. "I say, 'Do you want them back?'"

Henry thinks what really ends a mouse plague is not poison or traps. It's other mice. As populations get larger and larger, the animals just can't get away from each other. They start spreading disease, and even though they are feeding on the piles of leftover grain, eventually that grain runs dry. Mice get stressed, mice get hungry, and mice start eating each other. First the sick and weak ones, then the males go for newborn pups, slaughtered while the mother is out searching for food. Eventually the population crashes back down.

"I'm only guessing here," Henry says. "We haven't done the science to prove this." But he does know that when densities of mice are low, during dry seasons, there's a pretty high rate of infant survival. Males can't find nests of other mice to eat. When densities get incredibly high, though, it may be another matter.

Henry and his colleagues at CSIRO are still working to predict mouse plagues and trying to convince farmers to keep a close eye on the rodent populations and on their own waste. Every bit of grain that drops between a header and a combine during harvest is a mouse snack in waiting. It doesn't help that farmers are now embracing no-till farming, where seeds get sown right on top of the last crop. In dry or windy areas where there's potential for a lot of erosion (like Australia), the roots of the old crop help hold the dirt in place, fertilize the soil, and hold moisture. No-till farming is great. But if no one plows, no one disrupts the burrows of the mice that have taken up residence under the crop.

Then, of course, there's climate change. Weird weather tends to favor mice, Henry notes. And climate change offers plenty of weird weather. "I think that conditions like we've had this year, where [we've] had a lot of summer rainfall, actually helped to sustain breeding through those summer months," he says. Mice keep breeding through summer and fall, as long as they have plenty of food. Then, with a mild winter—more likely to happen due to climate change—more mice will survive. Which means spring starts off with more mice, which then will breed as more food becomes available. You get the idea.

Poisoning leads to the stench of mass death. Traps can't possibly handle the surge. As long as conditions are favorable, mice will breed, which is what mice do best.

What if we could take advantage of that?

GREGOR MENDEL WANTED to work on mice. The nineteenth-century monk famous for his pea plants actually started his inheritance studies using rodents. The local bishop, however, was not a fan. Mice, you see, had sex. Right in their cages where anyone could see. A man of God should not be encouraging such licentious behavior, even if only in mice. Plant mating systems, on the other hand, are remarkably unsexy. So peas it had to be.

If you've taken a biology class you'll probably remember Mendel and his peas. Genes come in pairs. A pea could have two alleles (versions of a gene), one for wrinkles (w) and one for a round pea (W). That pea could mate with another with the same alleles, one for wrinkled (w), one for round (W). Each parent can pass on only one.

If what you're looking at is Mendelian genetics, each offspring has a 50 percent chance of getting either W or w. If each parent is Ww to begin with, roughly 50 percent of the offspring will also be Ww. Another 25 percent will be WW and the remaining 25 percent will be ww.

This becomes a problem if a scientist wants to use genetics for mouse pest control, as many scientists do. Even if you introduce an allele for pest control by releasing tons of horny mice to mate with the ones already there, spread would happen only half the time. What you need is an allele, or set of alleles, that gets passed on more often than that.

Such a thing exists, and it exists in mice. It's called the t haplotype and it exhibits what scientists call extra-Mendelian inheritance. It's not just one gene, notes Anna Lindholm, an evolutionary biologist at the University of Zurich who has looked long and hard at the t haplotype. About nine hundred genes are linked together on a piece of chromosome 17, and they get inherited all together as one unit. (That's what a haplotype is, a group of genes passed on together as a chunk.)

The t haplotype is a natural gene drive—a gene or set of genes passed on more often than Mendel's peas would predict. If a female mouse has the t haplotype, 50 percent of offspring get it, just like normal. But if a male mouse has one copy of a t haplotype, that haplotype will get passed on 95 percent of the time.

Gene drives win by fighting dirty. The t haplotype goes right for the mouse's nuts. When mice produce sperm, the cells have only half the normal complement of genes—all set to pair with the egg's genetic half on the other side. If the mouse in question has a t haplotype, Lindholm explains, 50 percent of the sperm should have the t and 50 percent shouldn't.

Then the haplotype strikes. The genes in it instruct the mouse's other cells to pump out poisonous chemicals, Lindholm explains. The chemicals act on a section of the wild-type genes (that do not have the t), seeping through the tiny cell bridges connecting the developing sperm, leaving ruin in their wake. The t haplotype has those same genes, but it also has alleles that are resistant to the poisonous products it's pumping out.

"So it's carrying the poison, and it's carrying its own antidote," she says. "But the poor wild-type spermatids that are developing, they only get the poison and are damaged while they're developing." The resulting wild-type sperm are hamstrung and can't swim as well. "As they're racing through the reproductive tract to get to the female egg, the ts are getting there first."

This seems like an excellent way to exercise genetic control of the mouse population. All scientists would have to do is pop in an extra gene linked to the t haplotype, which would make all t haplotype mice infertile in just a few generations. The gene spreads like wildfire, and your mouse problem is over.

"When you introduce a gene drive, you're forcing the genome of the organism to spread it no matter what, even if it has fitness costs," says Raul Medina, an entomologist at Texas A&M in College Station, who thinks a lot about genetic control of pests. "For me that's the most important innovation because we have been able to do genetic modifications before, but normally they get lost, because they are usually not good for their organism."

But there's a piece missing here. If the t haplotype is passed on 95 percent of the time, shouldn't every mouse on earth already have it? They don't, and for good reason.

The odds for inheriting the t haplotype may be good. But the goods are odd.

Lindholm has kept a wild mouse colony in a barn outside of Zurich since 2002 to study mice in a natural environment. Food and bedding are provided. The mice have no incentive to leave (cats, hawks, and owls caught on very quickly and there's usually a few waiting outside), and every reason to stay. They stay, they breed, and Lindholm and her colleagues keep track of the results. For a while, there was a t haplotype spreading, but it didn't spread far. "We had about thirty percent for a few years, and then it dropped off and went down to nearly nothing," she says. "And then in 2011, we had the last one. . . . It went extinct."

The problem is that if the t haplotype can't cheat, it can't win.

"Once a female mates with a male t carrier and a wild type, then there's strong sperm competition between the sperm from these two males. And the t sperm are totally crap," Lindholm explains. Scientists don't know why but, when given the choice, a female mouse won't even mate with a male t haplotype carrier. The t haplotype mice are also more likely to take their rejection and travel outside the barn—making them more likely to end up someone's snack. Many variants of the t haplotype even have a lethal recessive gene included. This means that anything that inherits two t haplotypes will never get born in the first place.

But don't count the gene drive out yet. You see, not all t haplotype versions are lethal to the mice that inherit two copies. John Godwin, a comparative reproductive and behavioral biologist at North Carolina State University in Raleigh, is interested in one of the less lethal versions. After all, he says, "the t haplotype has been around for more than two million years, and evolution hasn't beaten it yet." It may have gone extinct in Lindholm's population, but it has never disappeared entirely.

Godwin is a member of GBIRd, Genetic Biocontrol of Invasive Rodents, an international group of researchers that is looking at techniques such as gene drives to control mice, and eventually rats.

The nonlethal t haplotype offered GBIRd an opportunity. It could be linked to another gene—the male-determining SRY gene. The SRY gene causes extreme male bias in a population—so extreme that both males and females end up appearing as males. The idea is that scientists could link a t haplotype to an SRY gene in male mice and release it into the population. Any female those mice mated with would have only male-presenting mice, the population would crash, and game over.

But in a series of models, Godwin's colleagues showed it wasn't that simple. It would take up to six releases of t males to overcome the wild mouse males and the female mice who would really like other dating options. Scientists would have to keep releasing the t haplotype mice over and over until they so overwhelmed the wild males that the females would just end up settling.

Maybe a crappy gene drive is the gene drive we actually need. It will only pass down to one generation, and then it's done. It might not even get that far, because the t haplotype mice are so relentlessly unsexy. It most likely wouldn't escape an island, say, where it was being tested, run amok, and drive the entire species of mice to extinction.

Which is something that people do worry about when it comes to CRISPR.

CRISPR stands for clustered regularly interspaced short palindromic repeats, which probably doesn't mean much of anything to most people. Not only that, when scientists these days say "CRISPR" what they actually mean is "CRISPR/Cas9."

Cas9 is an enzyme that cuts DNA—whenever, wherever. Since you don't want Cas9 just running around with molecular scissors, CRISPR produces tiny bits of guide RNA that act as a map, telling Cas9 where to cut. CRISPR/Cas9 is not a gene drive, though, Medina explains. It's just the scissors.

After DNA is cut on one strand, something called homology-directed repair happens. This is the frantic gluing of clipped bits of DNA around whatever you cut out or added. Because cells conveniently have two copies of each gene, homology-directed repair can copy the untrimmed copy onto the cut one. Both strands will then have the same DNA sequence. If that DNA is in an egg or sperm, it will then pass on that same sequence to 100 percent of its offspring. In those offspring, CRISPR/Cas9 will again go to work on any sequences it recognizes, cutting and forcing the DNA to copy over and over. This is called genotype conversion, and it creates a gene drive. After the gene is cut and the clean gene pasted via homology-directed repair, the animals will pass on the same sequence every single time.

Unfortunately, this only really works for bacteria, yeast, and insects. Mammalian cells tend toward a different and far more lazy method of DNA repair. They simply glue trimmed ends together rather than copying the clean DNA to the damaged section. How can a scientist get a gene to copy if mammalian cells won't make any copies? Kim Cooper, a developmental evolutionary biologist at the University of California, San Diego, was up to the challenge.

It turned out to be a matter of timing. If Cooper and her lab made the CRISPR/Cas9 system cut just as eggs and sperm were being made, the cells would do homology-directed repair. Cooper and her lab got it to work (only in female mice), increasing how often a single allele was inherited from only 50 percent to 72 percent.

Of course, it's never going to end up being that simple. CRISPR has a map, but sometimes fails at reading directions, letting Cas9 cut in the wrong place. The DNA can also give CRISPR/Cas9 the slip. DNA is constantly mutating. If one of those tiny mutations means that CRISPR's guide RNA can no longer find the spot it needs, the system will stop being able to change the genome entirely.

So it could be ineffective. But too effective is also a problem. If scientists developed a gene drive that was coupled to an infertility gene, and if the gene drive was good enough, it could drive an entire species extinct. And sure, we want mice and rats gone from islands, but we don't want them extinct. Scientists at GBIRd and elsewhere are already teaming up with ethicists and social scientists to figure out how to control gene drives and how to release them with the consent of humans living near the target populations.

To keep it localized, Godwin and other scientists at GBIRd are thinking about training a CRISPR system to hunt DNA that is specific only to a tiny subpopulation of rodents. Most mice on islands come from one or two that escaped from a boat—which means the population is very inbred, with highly similar genes. Scientists could (in theory—all of this is in theory) target CRISPR to a particular gene in the mice that existed only on that island. The drive would then spread to those mice by identifying that gene. But if a mouse did escape via a scientist's suitcase and made a life for itself somewhere else, any mice it mated with off the island wouldn't have that inbred gene. CRISPR could not attach, and the gene drive would never take hold.

Cooper's system isn't a full gene drive anyway. It works only in females, and her group only tested it in one generation. To be a true gene drive, the system would need to be inherited across many generations of mice.

Nevertheless, when her paper was published, other scientists immediately started thinking about the potential for using a gene drive for pest control. Cooper, however, was not. She's a developmental evolutionary biologist at heart. She wants to understand things like how jerboas evolved to have such long legs. "If you want to try to recapitulate the phenotype and understand its genetics," she says, "then you need to make multiple changes in one animal and see how they interact with each other." Her goal was to make those many changes easier, instead of breeding mice over and over and over trying to get vanishingly rare gene combinations (giving lab mice "freakishly long legs" will merely be a side effect).

Cooper didn't want to get rid of mice. She developed her system to study mice better.

She's not alone. In fact, many of the most impressive achievements in biology have done the same thing. The end goal is understanding human life, the human body, human illness. But along the way, scientists (myself included) have spent a lot of time trying to better understand not humans, but mice. Optogenetics, genetic knockouts, even CRISPR itself, all these things are techniques that have been developed—and mostly used, so far—in the mouse.

The laboratory is the mouse's new niche. A niche that we have created, this time on purpose.

MICE BECAME THE preferred laboratory animal in the early 1900s. Scientists in the 1600s did use mice but tended to prefer larger animals such as dogs and sheep. A lot of what we know about the circulatory system and blood transfusions is thanks to dogs. At the time, many dogs were strays. Scientists could easily make the argument that those animals were unnecessary, unwanted, and even dangerous. Man's best friend could be put to better use in the lab, uncovering secrets of the body for the aid of mankind.

By the 1900s, though, research was scaling up. The scientific method was well established. Just one or two dogs did not an experiment make. If a scientist wanted to study, say, the hot new science of genetics, that researcher would need lots of animals to do it, animals that needed to be fed and housed and bred, all in exactly the same way to minimize any experimental artifacts.

"Do you like mice? Of course you don't," Clarence Cook Little wrote in *Scientific American* in 1935. Little was the founder of the Jackson Laboratory, now one of world's biggest purveyors of mice for scientific research. The mice, descendants of which are still used today, resulted from the efforts of Abbie Lathrop, whose detailed note-taking and breeding techniques produced the strains Little worked with in his early work at Harvard—and took to form the Jackson Laboratory. Now, Little had to sell his mice. And to sell mice, Little had to convince scientists to buy them.

Mice were pests, of course, Little wrote. But they could be something else. Little had already been "selling" the idea of genetically identical inbred mice to doctors and scientists across the country. Now, in his article in *Scientific American*, he began to make the idea palatable to the public. He offered a new role for mice, as "the troops which literally by the tens of thousands occupy posts on the firing line of investigation" into the "nature and cure of cancer."

In picking the mouse, Little landed on an animal enough like us to be scientifically worthwhile, and enough unlike us that we didn't feel too many pangs of conscience. Early animal rights activists (then called anti-vivisectionists) could easily get backing for stopping research on dogs, monkeys, or rabbits. But mice? That's a harder sell, says Karen Rader, a historian of science at Virginia Commonwealth University in Richmond and the author of *Making Mice: Standardizing Animals for American Biomedical Research, 1900–1955.*

Choosing an animal like a mouse—a pest that no one liked—had a lot of PR advantages. "I'm pretty sure Little understood that," Rader says.

Little's main interest was actually eugenics, with many strains of mice inbred to the point of having almost completely identical genomes. After World War II, the dangers of fixation on the perfect man—or perfect mouse—became all too clear. But the "purity" of Little's tiny furry clones also fulfilled another role in the lab. These mice were the living embodiment of a carefully controlled scientific experiment. With inbred strains of mice, Little was selling an idea of how science could be done.

Scientists bought, and the public listened. Jackson Laboratory, known to the research world as JAX, now keeps twelve thousand strains of mice and sells more than three million mice per year.

For the first time, standardized lab procedures could have a standardized mouse. "Having a standardized mouse was Little's obsession," Rader says. "That obsession reshaped the history of medicine."

Scientists have developed methods to study pretty much everything in mice. Numerous new treatments for everything from anthrax to Zika have passed through their tiny bodies.

Antidepressants, cancer treatments, diabetes, Alzheimer's, Parkinson's. Little's vision has come to pass in many ways. Mice really are heroes of medicine.

FOR THOUSANDS OF years, we've spent time trying to get rid of mice. But now, modern medicine has taken something no one wanted and made it very valuable indeed. We have taken it, bred it, changed its genes. We have made a new niche for the mouse in the laboratory.

In both cases, we played an active role. We created an environment where a brave mouse could thrive. We made the new, sterile laboratory niche where mice are now passing along identical genes by the million. Consciously or not, we made the mouse what it is today.

The mice aren't innocent bystanders either. They got those grains. They adapted, and they are still adapting, to live among us better than ever before. We engineered a niche, and they evolved to fill it.

Mice as a pest have succeeded wildly. Now as a laboratory model, in a way, they are still winning. After all, if the goal of an organism is to pass on its genes, millions of mice do it in labs around the world every day.

4

A DROPPING
OF PIGEONS

At Harvard Square in Cambridge, Massachusetts, Ivy League–seeking tourists and crowds of undergraduates stream daily past a woman. She usually huddles on a milk crate, across from the entrance to the CVS, the hood of her black puffy jacket pulled tight around her face. Her belongings, sheltered by colorful and neatly fitted covers, sit protected in an unused doorway. Her sign says, GOT EMPATHY?

Her name is Carrie, and it's easy to get the impression that everyone in Cambridge knows her name. Tourists pass her by without making eye contact, with the embarrassed look of people who feel guilt, but not quite enough guilt to respond to her request for empathy. They have no idea they're ignoring a Harvard Square institution. She's cheerful, friendly, and like many of the people around Harvard Square, extremely well read.

Sometimes she sits silently, closed in on herself, trying to make a smaller target against the cold. Other times she chats animatedly to one of her many friends—residents, other people who spend their days in the square, students, and local cops. And whenever she can, she feeds the pigeons. She feeds them granola, when she can get it, but when she can't anything will do. KIND brand granola, she says, is their favorite.

I came upon her one day as she stood in a flurry of wings, smiling hugely. Two pigeons perched on her gloved hands, pecking gently at the Frosted Flakes cupped in her palms. One had a patch of white feathers on his tail. She had named him Niels Bohr, after the famous physicist. Her delight in the birds on her hands seemed to make the whole square a little sunnier. Suddenly, she looked up and anxiously scanned the crowd of pigeons. "I've got another, he's usually around," she says. "Where is he? I call him Fyodor Dostoyevsky."

When she's huddled in on herself, sometimes she's actually snuggling a pigeon.

Some people walk past her, see Bohr and Dostoyevsky on her hands, and smile. Others grimace, dodging and weaving in the feathered crowds, trying to avoid stepping on one. A few don't even dodge, they just walk right through, and pigeons get out of the way or suffer the consequences.

Carrie's bird-feeding habits aren't always popular. "One lady," she says, "has been calling Animal Control three times a week. She wants to have me arrested. She says I'm a danger to the birds—I'm feeding them too close to the road."

When I speak to my friends and colleagues about pigeons, many people respond with simple disgust at the idea that anyone would feed pigeons. Over and over I heard that pigeons were dirty, that they spread disease.

And when I tell them about Carrie, they get even more uncomfortable. Though they can't say why, precisely. When he went to write his dissertation on parks in Greenwich Village, Colin Jerolmack—now an environmental sociologist at New York University—thought he would study how humans developed affection for urban spaces. But after hanging out at parks and town council meetings, he realized something else. "The two big problems that people talked about incessantly—and they talked about them in the same way—were pigeons and homeless people," he says. Perhaps people feel discomfort because they're forced to see something they don't want to see. They don't want to see evidence of people in poverty, intruding on their comfortable lives. They don't want to see pests—animals that aren't wild, but aren't accepted either. Society has, in a way, failed them both. The pigeons thrive in spite of it. And because of the pigeons, Carrie smiles in spite of it.

To Carrie, the pigeons aren't a pest. They bring her joy. And they used to bring joy—or at least, calm acceptance—to people all over the world. Humans created a place for the pigeon, bringing it willingly into their lives.

This is a species that helped to found modern journalism, a species whose members received military awards for bravery. And now, they're "rats with wings." What changed?

THE RULE OF who gets domesticated is the same as the central rule of real estate. Location, location, location. The pigeon, *Columba livia*, is technically also the "rock dove," or "rock pigeon." Fossil evidence suggests that the original range of the rock dove might have been between the Middle East and South Asia. That huge swath includes the Fertile Crescent, prime real estate for domestication. Pigeons love nesting in cliff faces, leaving home daily to forage on grains and seeds and returning home at night. When humans in the Fertile Crescent started growing lots of grains, all grouped together in tidy plots? Pigeon jackpot. Even before humans started domesticating the pigeon on purpose, it was probably following people and their farming around, says Will Smith, a zoologist at the University of Oxford who is studying feral and wild pigeon mixing.

This was a golden opportunity for early farmers. These were birds that went out, fed themselves, and came back every night to exactly the same place. Birds that went on to produce more birds and could do so all year under the right conditions. Birds that liked cliffs—but were happy to accept buildings instead.

Pigeons became one of the earliest domesticated birds. Scientists estimate that humans were keeping rock doves five thousand years ago, if not earlier. In the *Epic of Gilgamesh*'s version of the flood myth, a pigeon is the messenger. Pigeons are sacrifices in the Old Testament, and in the Gospel of Luke, Jesus is consecrated in the temple, along with a sacrifice of two turtledoves, or two pigeons.

When humans developed villages and then cities, pigeons came along for the ride. "In rabbinic literature, pigeons are very much valued," explains Beth Berkowitz, a Judaic scholar at Barnard College in New York City. "[Rabbis had] laws about who owns a pigeon . . . they were a valued property like sheep or goats." Pigeons were so much a part of human life, she says, that you couldn't even capture your own near a settled area. "You can only take a pigeon from the wild if it's a certain number of miles from the town," Berkowitz explains. "If it's close enough to the town, you can't take it because it's seen as someone's property." A city pigeon wasn't an interloper, it belonged to someone.

People willingly brought pigeons over to the Americas. We hunted them for sport, ate them, competed them in races. We began to breed them for their looks—"The pigeon fancy." Nikola Tesla fell in love with one, and Darwin devoted a whole chapter of *On the Origin of Species* to the pigeon fancy. "His editor encouraged him to just drop all that stuff about natural selection, and just make it a pigeon book," says Colin Jerolmack.

Yet, there were loads of escapees. Pigeons took to the streets, giving rise to the flocks of feral pigeons we see today. As urban populations increase, feral pigeon populations do too. And the more common they get, the less we like them. "We didn't really have negative feelings about them until the twentieth century," Jerolmack says. In fact, he can tell you the very day and year that someone in New York City first published the term "rats with wings": June 22, 1966. He analyzed every single mention of pigeons in the *New York Times* from 1851 to 2006. The "rats with wings" comment came from Parks Commissioner Thomas Hoving.

It was a real about-face. The first few articles about pigeons in the 1870s and 1880s were bewailing the people who shot them, claiming the hunters were committing "brutal murder." It was sparrows people hated, not pigeons. Sparrows were "lazy" and "audacious." Pigeons were noble and innocent.

But the tide was turning. In 1927, the director of the New York Public Library begged people to stop feeding the pigeons. People were banned from keeping pigeons on the roofs of New York City in 1930. By the late 1940s, people were done with pigeons. They put out poison and accused the birds of spreading disease.

What caused this? What turned the pigeons from beautiful, useful birds to pestilential pest? In part, pigeons can lay blame on chickens.

Before the rise of the chicken on the American plate, city dwellers ate pigeons, Jerolmack says. "If people still wanted to actually kill their own food that they kept in their backyard or on a roof, pigeons are great." They live in your dovecote, then they go out during the day and feed themselves. They come back at night, breed, and you eat the proceeds—tender, fluffy squab.

As writer Courtney Humphries documents in her book *Superdove*, squab started out as a delicacy in the dovecotes of the rich. High prices on pigeons in the early twentieth century promised profit to American pigeon breeders. But in the Great Depression, pigeon prices tanked, making squab more of a food for the people. By

1944, a member of the U.S. Department of Agriculture estimated pigeon plants were pumping out a hundred thousand birds a year. And that doesn't count the many city and country dwellers who kept pigeon lofts on their roofs—self-feeding, self-cloning protein.

Pigeons were profitable in other ways too. As speedy fliers that rarely lose their way, pigeons were used as messenger services in ancient Persia. Even modern journalism owes a debt of gratitude to the pigeon. Paul Reuter, the founder of the famous wire service Reuters, started out flying hot stock tips between Brussels and Aachen in the late 1840s, beating the train by two whole hours. The birds also flew home from the fronts in World Wars I and II carrying vital information on troop movements. They even earned medals for valor.

By the end of World War II, though, pigeons weren't so useful anymore, Jerolmack says. The heroic messenger pigeon was replaced by the rising availability of the telegraph, the telephone, and then the cell phone. Squab was replaced with chicken. "Animal husbandry disappears," he says. "Nobody's keeping any kind of livestock." Instead, they're buying meat at the butcher, and then at the grocery store.

Pigeons flew higher, further, and faster. But poultry became about which birds grew bigger, fatter, and better. "So then it's about economies of scale," Jerolmack says. "And pigeons don't scale up well."

Having purchased, cooked, and eaten squab with urban ecologist Elizabeth Carlen, it's possibly also a matter of taste. Pigeons and chicken both have breasts. But it's pretty clear who's built to pump their wings and fly. The pigeon meat is dark and rich, the muscle of an endurance athlete, while the chicken's soft white breast is bred for a life of flightless leisure. The pigeon was gamey and slightly smoky. At first, it was foreign, but after a while, Carlen was cheerfully picking at the carcass. By contrast, the paler meat of a similar-size Cornish hen was flat, flavorless. It tasted like chicken.

Humanity has a short memory. And what has the pigeon done for us lately? Pigeons had purposes, but they outlived them all. They stopped perching and began to loaf. They became what Jerolmack called pedestrian animals. "We see them every day, but they're also literally on our sidewalks."

The history of pigeons has something else to show us too. City pigeons aren't wild. The pigeons strutting the streets today are descendants of those that lived in dovecotes, the great-great-great-grandbirdbaby of something people valued very much. Until we didn't. Until we let them go, declared them useless, and decided they were just rats with wings.

The pigeon on the street was still a success in terms of animal adaptability. Doves and pigeons in general have good potential for living near people, says Louis Lefebvre, an animal behaviorist at McGill University in Montreal. Eurasian collared doves, Lefebvre points out, also have become city dwellers and are an invasive species in North America. "Columbiforms [that's pigeons and doves] are more urbanized than we would predict," Lefebvre says. "What makes pigeons and doves so good at being urbanized? It probably isn't brain size . . . if you compare them to all birds, they're not very large brained."

Instead, it has to do with being a generalist. Like the mouse or the rat, pigeons don't mind living in human structures and eating human foods. If you can live in lots of places (like cliffs or buildings), and eat lots of things (like seeds or pizza crust), you've got a leg (or wing) up on the competition in an urban environment. "There's nothing special about being an urban dweller or being able to live in the city, you have to be a generalist, and you have to be probably a bit tamer than other species," Lefebvre says.

We have also cultivated skills in pigeons that serve them well in our environment. "The way that pigeons were submitted to artificial selection is like the best of both worlds," he says. Pigeons were selected to be so tolerant of humans that they'd respond to a person taking away their nestling by just popping out another egg. But at the same time, unlike chickens or doves, pigeons weren't fed. They were expected to go out and feed themselves.

Feeding themselves, Lefebvre says, kept pigeons streetwise. Many birds—like the ring dove or African collared dove (*Streptopelia roseogrisea*), domesticated in, er, Africa—were domesticated to be tolerant of people. But "when we tried working with ring doves, we couldn't get them to do anything," Lefebvre says. "Because they just look at us with their cute little face and say, 'What do you mean, I'm not getting free food? You want me to work and learn something? Are you crazy?'" Pigeons persist because working for food is something they're used to. Lefebvre has shown that urban birds are better at problem solving, like house mice or suburban raccoons.

We created the urban environment. We also created the pigeon in its current form—the tame, speedy, fast reproducer that's not a picky eater. Is it any wonder they did so well?

EUROPEAN RABBITS (*Oryctolagus cuniculus*), like pigeons, are also a self-feeding food source. And of course, they have the benefit of breeding like . . . I don't need to finish this sentence, do I? This made them great for European seafarers, who spread both rabbits and colonialism. "They just stopped on an island and chucked a couple of rabbits on, so when they came back six months later, they had a protein source that was super easy to harvest," says Tanja Strive, a scientist studying rabbit biocontrol at the CSIRO in Canberra, Australia.

One might think that this strategy—chuck rabbits on island, come back for free dinner—is what introduced the rabbit to Australia. One would be wrong. Rabbits were brought to Australia by a man named Thomas Austin. Austin was a member of the Acclimatisation Society of Victoria, a club of European colonizers interested in the economic prospects of Australia but terribly disappointed by the wildlife.

I mean yes, Australia had kangaroos and kookaburras and dingoes and koalas and wombats. But it all looked so foreign. It was nothing like England. And if you were an avid hunter like Austin, Australia was really lacking in one thing—rabbits to aim your gun at. So in 1859, the ship *Lightning* arrived with a special bunny delivery. Some say it was thirteen rabbits and others say twenty-four. No matter what, Austin put the rabbits in large enclosures to keep them in one place so they could breed and he could hunt the proceeds. "It was a very gentlemanly thing to do back in the day," Strive says. The rabbits, however, escaped. Of course they did.

Rabbits liked Australia. There was enough to eat, not too many predators, and the soil was nice to dig in. Rabbits did what rabbits do. They did it so well that Australia was soon completely overrun with bunnies.

In fact, in Australia, Strive notes, rabbits contributed to the effects of the Great Depression. Like mice, rabbits can reproduce to plague proportions, and in the 1920s and 1930s they did. "They basically destroyed agriculture, they destroyed arable lands, people had to walk off their farms, because there was nothing left," she explains. "What little vegetation was there was denuded by rabbits, and then the topsoil went off," blown away without any roots to hold it down.

The only possible silver lining was that rabbits were meat. During the Great Depression, Strive says, rabbit was on a lot of menus. The army even prepared tinned rabbit as rations for soldiers in World War II. The skins also proved valuable. They make nice hats.

But the war ended. The Depression ended. Any use for the rabbits—hunting or food source—was gone. The rabbits remained. Australians built huge rabbit-proof fences. They poisoned them, hunted them, and made more rabbit-skin hats. But no hunting season, hat industry, or poison was enough.

Like pigeons, rabbits had traits that people valued. Speed-breeding meat, fur, target practice. But their very breeding abilities—the thing that made them useful—also made them a pest. In desperation, scientists raced to find a disease that would kill bunnies. Faced with a plague of rabbits, they began to look for a rabbit plague.

PIGEONS STILL RETAIN the traits we used to value—they race home with unerring speed, they flutter through the air with tasty meat. And they are still domesticates. It's a glorious spring day and I'm out with Elizabeth Carlen, the urban ecologist and postdoc at Washington University in St. Louis, Missouri, and Jonathan Richardson, an urban ecologist at the University of Richmond in Virginia, at Byrd Park in Richmond, accompanied by a gaggle of University of Richmond undergraduates. The area by the fountain glows in the afternoon sun. It's filled with picnickers, geese, and pigeons.

Carlen wrote her dissertation tracking the genetic relationships between the members of the giant mega pigeon flocks of the East Coast. Now, she's giving a demonstration of her techniques to Richardson's students and one extremely enthusiastic journalist.

Carlen and Richardson stride around casually, tossing out bird feed with practiced swings, while the college students hang back, unsure what to expect. Once the pigeons are in a huddle around a particularly thick patch of feed, Carlen steps confidently forward, aiming in front of her with what looks like a large black flashlight.

There's a popping sound, a bit like an air gun, and the flock takes off, leaving six of their number fluttering uselessly against a large net. Carlen deftly scoops up pigeons and net together, holding the birds carefully against her torso, and heads back toward the now clamoring students. She sets her catch on the ground and begins easily unthreading them from the net, bird-butt first. As each one comes out, she passes it off, focusing intently on the next.

The students, one by one, hold out their hands for pigeons, some with excitement and some with extreme trepidation. But once they are holding a bird, their confidence soars. The pigeons are shockingly docile. Soon, the students are grinning and cooing. The pigeons quickly receive names—Milo, Daphne, Frederick, Georgia, Alfie, and Billy—even though it's impossible to tell if you're looking at a female or male pigeon without a thorough dissection. When Carlen begins weighing, banding, and taking a small blood sample from each bird, the students cheer on their charges with proprietary pride. When Milo weighs in at a healthy 450 grams, his temporary handler grins indulgently. "That's my baby!"

After some weighing and banding, a student passes me "Daphne." The pigeon fits perfectly in my hands. Its iridescent neck and throat shine, and its orange eye glances at me with slightly ruffled patience. There's not a single hint of aggression. Hold a pigeon and it sits perfectly still, except for dropping a green turd once in a while. "That"—Carlen grins—"is [thousands of] years of domestication." A pigeon—even if it's never been held in its life—is, somewhere deep down in its makeup, used to being held. And even though I've never held a pigeon in my life, my hands seem instantly used to holding it.

It is rather like holding a large tense chicken breast.

The birds seem clean enough. At least, they don't have dirt on them. But as I hand off pigeons, and help to take blood samples and notes, I begin to notice tiny bugs. One on a wing. One on my wrist. One on my notebook. Later, over dinner, Carlen laughs. "I didn't want to tell the students, but those pigeons were really lousy." My neck instantly begins to itch.

As I drive home, hours later, I feel the itch creep down my neck, over my arm. It's definitely my imagination. Carlen tells me later that pigeon mites don't generally cling easily to featherless human skin, a fact I confirm with a scientist who studies pigeon lice. But in the car, I just can't shake the feeling. When I finally burst through the front door, I gasp to my partner, "I'm sorry I'm late! Everything I'm wearing needs to go in the washer right now. I'm covered in lice."

I'm not the only one who worried about dirty birds. In 1945, Philadelphia killed hundreds of pigeons because they were infected with "ornithosis." In 1952, scientists confirmed that yes, pigeons could carry the disease (though by that decade it was also called psittacosis). It's a bacterial infection caused by *Chlamydia psittaci* (yes, related to *that* chlamydia), and results in fever, dry cough, muscle aches, and pneumonia—in humans. The birds do not report any symptoms—but humans observing them see poor appetite, diarrhea, and inflamed eyes.

By 1961, pigeons were officially labeled a health menace in the *New York Times*. By 1963, signs in Queens said DO NOT FEED PIGEONS. PIGEONS ARE THE GREATEST DISEASE CARRIERS.

As noted, pigeons can carry psittacosis. But it's a rare infection, and cases are more common in people who spend a lot of time around other birds such as parrots, or farm birds like chickens or ducks. You have to be close to a bird, and more important, to its poo, where the bacteria thrives. "Yes, it is theoretically possible" that people could get it, Jerolmack says. "For instance, if you went into an abandoned building, where for years pigeons had been nesting and defecating, and nobody was cleaning up that feces, and you hung out there and breathed it in? Could you get sick? Yes. But are you going to get sick from even sitting on a bench with pigeon shit? Or breathing in pigeon crap in an outdoor space? Or even if there was one in your attic? No."

More worrisome pigeon-carried illnesses are fungal diseases like cryptococcosis, caused by *Cryptococcus neoformans*, says Anastasia Litvintseva, a mycologist at the Centers for Disease Control and Prevention in Atlanta. *C. neoformans* is a single-celled fungus that lives pretty much everywhere in the environment. Most of the time, we'd never notice, but for people with weak immune systems—such as people suffering from HIV/AIDS—*C. neoformans* can be deadly, causing a type of meningitis.

While people might get sick from *C. neoformans*, pigeons usually don't. They run hot, with a body temperature of around 42 degrees Celsius, Litvintseva explains. It's too much for a delicate fungus. But like psittacosis, it's not the pigeon that matters, it's the pigeon poo. "It's a great growth medium, it's like a petri dish," she says. In fact, Litvintseva has shown that pigeons might have been one of the causes of the most common strains of *C. neoformans* spreading around the globe. "They're probably not the cleanest birds in the world," she says. "They might have, you know, carried it on their feet or feathers."

C. neoformans is not the only fungi around that's no fun at all. Another fungal disease, histoplasmosis (courtesy of the fungus *Histoplasma*), loves a meal of pigeon poo. It's a soil fungus, explains Tom Chiller, another mycologist at the CDC, but bird and bat poo is too good an opportunity for histo to pass up. Again, it's generally a problem for people with weakened immune systems. For others, their worst histo infection, Chiller says, "feels like you got a cold. It goes away, [and] we never knew you had it." But histoplasmosis, like psittacosis or *C. neoformans*, isn't specific to pigeons. "When I think of histo from a public health standpoint, I don't really think of pigeons," Chiller says. "I think of birds and bird roosts. It's a matter of nitrogen rich soil, so we think pigeons because they clump together and poop a lot, but a lot of birds do that."

Unless we're in the pigeon fancy, most of us aren't going into pigeon-filled lofts with lots of poo. But in urban areas, pigeon populations get quite high. If the birds crowd (and poop) around hospitals, for example, when that poo dries out and floats into the air, it could present a risk to immune-compromised patients, Litvintseva says.

If this sounds terrifying, well, yes. But pigeons are similar to other domesticated and pest animals in that respect. Cows, pigs, chickens, turkeys, mice, rats, dogs, and cats can all give diseases to their un- witting human neighbors. Like pigeons, they can pass on bacteria, viruses, and fungi without ever getting infected themselves—as we scoop and inhale their poo. When people brought other species close to us, that was a risk we took on. We reap the benefits—very good dogs, cats on laps, ice cream, and bacon—but it comes at a cost.

People's concern about pigeon diseases is mostly about what pi- geons might give to humans, not the bird's health. But when Tanja Strive and her Australian colleagues wanted to stop the rabbit menace in Australia, they also turned to disease. And this time, they looked for something, anything, that would make a rabbit very, very sick.

BY THE EARLY 1900s, approximately 70 percent of Australia was playing host to rabbits. This was no longer about seeing lots of bunnies. Their voracious appetite for greens killed off the grasses that held soil in place. Dirt dried up and blew away from people's farms, leaving only rabbit burrows in its wake. The only exceptions were the very driest desert and very wettest tropical areas. Australia is about the same size as the lower forty-eight U.S. states (maps don't do it justice). Shooting and poisoning, eating and making rab- bit hats were not doing the trick. And so scientists started looking around for what they like to call biocontrol.

Previously, biocontrol meant bringing in a natural predator to eat your pest. Australia did try this for rabbits, releasing cats and mongooses into the bush in the late 1800s. The mongooses couldn't handle the climate, and rabbit trappers also trapped mongooses— trying to protect their business. The cats, however, did very well for themselves. Unfortunately, as Australia has since discovered, cats do not limit their diet to rabbits. Cats became a pest in their own right (more on that later), and the bunny population hopped along.

By 1887, New South Wales was offering a prize of twenty-five thousand pounds to anyone who could find a method to kill bunnies. "Actually, Louis Pasteur had a crack," Strive says. He sent his nephew to try inoculating rabbits with chicken cholera. (It was a failure.)

Luck struck, finally, in Uruguay in 1896, where scientists described a new, horrible disease killing imported European rabbits— myxomatosis, caused by the *Myxoma* virus. The disease probably jumped to European bunnies from the local forest rabbit, or tapeti, where it wasn't nearly as nasty. In European rabbits, though, it was devastating.

Australia began testing the virus on closed populations of rabbits. *Myxoma* spreads on the mouthparts of flies, ticks, and mosquitoes. But in the dry climate of Australia, there often weren't enough biting bugs to really move it around. Until 1950, when scientists began testing along the Murray River. The summer brought large numbers of mosquitoes—and *Myxoma* took off.

At first, it was a rousing success. The rabbit population took a nosedive from more than six hundred million to one hundred million. But natural selection will always come to get you. *Myxoma* evolved to become slightly less virulent. Where it had killed quickly before, it began to take weeks. Weeks in which rabbits developed huge swellings on their ears and eyes, a PR and humane disaster. Rabbits also began to develop resistance, and some never got sick at all. By the 1990s, the population in some areas was back up to 25 percent of what it had been before.

Time to find a new virus, and one was already on the way. In 1984, China found a hemorrhagic disease in rabbits imported from Germany. It's often called rabbit hemorrhagic disease virus, or calicivirus (pronounced like the "Khaleesi, mother of dragons" + virus). It was new, and it was unimaginably deadly. Australia pounced.

They imported the virus in 1991 for tests. The disease was fatal and fast. "Within thirty-six to seventy-two hours after the infection, the rabbit is dead," Strive says. The animal will only run a fever or show it's sick at all for about twelve hours before its demise. The virus can enter through any orifice, though usually it's through the mouth, and appears in the feces, which is lucky, as rabbits love to eat each other's poo. Calicivirus remains active in the dead rabbits for up to months after death, making it resurge again and again.

In 1995 tests moved to Wardang Island, where rabbits had previously been isolated when scientists were testing *Myxoma*. This time, while the bunnies were isolated, insects were not. The virus escaped to the mainland. In a show of extreme irony, the government tried to isolate the virus again—the virus they were testing to see if it killed rabbits—by killing rabbits around the outbreak.

It was too late. Calicivirus proved to be an even better killer than *Myxoma*, knocking back populations by up to 95 percent. To make lemonade from lemons, the virus was registered as official pest control in 1996.

Releasing a rabbit plague o'er the land was another instance of utilitarian ideas of what is right, says Anne Quain, a veterinary ethicist at the University of Sydney. Utilitarianism (like the utilitarian view of the landscape discussed in the introduction) is basically trying to do the most good for the greatest number of people. People, not animals. "Utilitarian arguments are not framed in favor of animals," Quain says. "Because when it comes to costs and benefits, they bear such high costs compared to us. It would never be seen as humane to release a disease that affected humans in that way." It helps, too, that most of the time, rabbits aren't dying where people can see them. "If you don't see the cost, it didn't really happen," Quain adds.

Calicivirus also had a limited life span. Rabbits began to adapt, and young rabbits became able to survive infection and come out with resistance. Then in 2015, rabbit populations got infected with another, deadlier strain that arrived accidentally from France. Rabbits died again, started becoming resistant again, and the population is again bouncing back. Australia is in an arms race, deadly disease versus bunny. "It doesn't really matter what you throw at rabbits, eventually they'll find a way to come back," Strive says. Like the Red Queen in *Through the Looking-Glass*, Australia's scientists are running as fast as they can, just to stay in the same place.

PIGEONS HAD ONE more useful thing to give—their poo. Depending on the size, age, and diet of the bird, a pigeon can poo between nine and twenty-eight pounds of waste per year. Yes, that's wildly variable. But so is a pigeon diet, says Maggie Watson, a conservation biologist who has studied pigeon poop pollution at Charles Sturt University in Australia. High-quality grain or birdseed would produce some copious crap. Wild birds eating grass seed would perhaps excrete something a little more delicate. But given that a pigeon weighs only between nine and thirteen ounces, and is only about eleven to fourteen inches long, it would always be theoretically possible to bury the bird in its own bowel movements.

In ancient times, pigeon guano was its own product. It was used to tan leather and fertilize crops. It's even a source of saltpeter for gunpowder (where it provides oxygen for the explosive reaction that propels the bullet). But modern chemistry made bird guano obsolete by creating cheap, plentiful fertilizer and synthetic nitrates for better guns.

When they do the doo on our fancy buildings, conservation biologists no longer cheer for free fertilizer. Instead, they describe the effect in their scientific papers as "aesthetically unpleasing." It's unscientific, I suppose, to say "gross."

It's in this aesthetically unpleasing morass that fungi and bacteria make their home. But that's not what building owners or guardians of historic statuary are worried about. What they're worried about is the corrosive effect of the poop itself. Many historic and elaborately carved buildings, made of sandstone and limestone, form lovely cliff-like structures where pigeons love to perch. When anything even remotely acidic hits, the stone will begin to degrade. It's rock, so it takes a while, but this is why acid rain and air pollution are so dangerous to the historical tourist draws of Sacré-Coeur and Trafalgar Square.

A pigeon doesn't have an anus, rectum, and vagina or seminal duct like mammals do. Instead, waste, eggs, and sperm all come out through a single all-purpose hole called the cloaca. Disgusting, yes, but undeniably efficient. This means bird excreta contains uric acid, a chemical in urine. It's slightly acidic. But this can be balanced out by other things in the bird's diet. Maggie Watson and her colleagues have shown that pigeon crap runs the gamut from a slightly acidic pH of 5 to a slightly basic pH of 7.4.

As an acid, pigeon poo at its worst isn't much worse than acid rain. Some studies have shown pigeon poo—mixed into delightful slurries or pastes and left for a good long time—might degrade copper or bronze. Unimpressed with these findings, "we decided to collect some poo and put it on there and see what happens," Watson says. After twenty days of observing pigeon poop on various stone tiles, "nothing really happened." A few stains, but that's it. In a later study, she even found historic conservators didn't really care much about pigeons. On a scale of 0 to 10, birds scored about a 5 in terms of how big a problem they might be, far below things like dampness and potential earthquakes, and just above vandalism. The whole idea of pigeons degrading masonry, she found, was something people just said without any evidence. A load of pigeon poo.

Regardless of harm to the physical buildings, when pigeons flock in hundreds or thousands, they can coat the surrounding areas in slimy crap. While the conservators might shrug in resignation and grab the hose, their bosses tend to complain. So it's no surprise that pest control professionals have come up with new and varied methods to try to get pigeons to leave.

Some people might try bird alarm calls, which could work on birds like starlings or crows. But that's no good for pigeons, says Jack Wagner. He's the owner of BirdBusters, a pest management company devoted to defeating the feathered menace. Audio efforts like bird calls are especially useless, as pigeons don't have distress sounds. Ultrasonic tones are even worse. The high pitches might irritate local teenagers, but pigeons can't even hear them.

"Exclusion is the solution," Wagner says. For example, pest control operators often spread tight mesh nets across delicate masonry. Stretch it tight enough, and people can see through it to the pretty columns and statues they've come to gawk at. And pigeons bounce right off—for the few years before you have to replace the net, anyway.

Some people try nasty metal spikes or coils on the artful architecture to keep the pigeons out. But a pigeon in a pinch is a creative bird. They'll pile trash, sticks, and grass well up on the spikes or coils, until they can land anyway. A more entertaining option is the slide, which is simply a sheet of metal that transforms all flat surfaces into a surface too angled for a pigeon to land. If a building manager's rage turns to violence, they might try electrifying wires to zap any bold bird.

The "latest and greatest" in pigeon management, Wagner says, is a little less painful and a little more amusing. "BirdFire" admittedly doesn't sound very nice. But it's all about the illusion. When placed on a ledge, BirdFire looks like little plastic dishes filled with yellow goo. Not very impressive, especially when perched on a stone cupid's head.

Pigeons can't hear in ultrasonic. But they can see in ultraviolet. The gel glows in the ultraviolet range, producing the illusion of fire. A pigeon coming in for a landing will see what appears to be a wall of ultraviolet flames in his favorite spot. It may mean you have to tape little plastic dishes all over your decorative stonework at almost fifty dollars for a pack of fifteen. But to keep the birds off, it might be worth it. Until the birds realize the fire packs no heat. Personally? I give it a week.

EVERYTHING HAS BECOME disposable. Sometimes that's because it's designed to be cheap. And sometimes it's planned obsolescence. When new smartphones come out with faster speeds, more storage, and even better photo filters, we race to embrace them. Months or years later we find our old phones and snort at the useless piece of junk. We wonder how we ever dealt with the crappy battery life and the actual, physical buttons. We toss it out with no hesitation.

Animals aren't so easily thrown away when they become obsolete. Pigeons plummeted from grace and became rats with wings. But they aren't gone. They're still here, long after we forgot what life was like with them. We see them as useless, animals whose only niche now is in the human ecosystem.

They could recover, though. Because pigeons are not the only animals to lose their place in our affection. Dogs—yes, dogs—have gone through similar ups and downs. In much of the world, though, they've regained their use, and our love for them has multiplied.

For example, many modern Muslims have pampered pet dogs, just like many non-Muslims. But for a long time, there was a pervasive stereotype that Muslims thought dogs are "ritually impure, that they can void someone's prayer, that their saliva is dirty, things like that," says Alan Mikhail. "You shouldn't keep them in the house." Mikhail is the chair of the history department at Yale University in New Haven. He's the author of a number of books on Middle Eastern history, and is especially interested in nature, animals, and environment.

In his book *The Animal in Ottoman Egypt*, Mikhail tracked down the history of the impure, unclean canine. And he found that, like many topics in religion, the negative view of dogs has a lot more to say about culture than it does about religion.

Muslims are not only following the Qur'an but also the Hadīth of the Prophet Muhammad. The Hadīth (which translates to "the story" or "the discourses") are "accounts by people who knew the Prophet," Mikhail explains. "Who talked about what he did, what he said, how he comported himself in all manner of things." It is in the Hadīth where you find most of the recounting of the Prophet Muhammad's dealings with dogs.

And there are Hadīth that say a dog's saliva is impure. But Mikhail notes that lots of Hadīth don't agree with each other. For example, there are other accounts that say dogs entered freely into the Prophet's mosque and lounged around at their ease. There is also plenty of emphasis on treating animals—however pure or impure they might be—with kindness and compassion. In Islam, animals and humans are worthy of the same consideration; cruel things done to an animal are just as bad as cruel things done to a person.

Pure or impure, in the Ottoman Empire, dogs had an important, protected place. The dog's place wasn't necessarily as beloved companions. No, most dogs did something that most still enthusiastically do today: they ate garbage. Dogs were the waste disposal system of Cairo in the time of the Ottomans. They were so good at it that visitors to Cairo marveled at the large packs of dogs roaming on seemingly every street.

The Ottoman Empire conquered Egypt in 1517 and ruled it, with one small exception, until 1867. That one small exception was all five feet six inches of Napoleon Bonaparte. In one of Napoleon's earlier land grabs on behalf of the French Directorate, he invaded Egypt, and the French occupied the region until 1801. In the process, they had time to discover the Rosetta Stone, dig up evidence of a previous Suez Canal that linked the Red Sea and the Nile, and carry out a mass poisoning of Cairo's dog population.

It's not that the French hate dogs. It's that these dogs had outlived their purpose.

"The overarching reason is colonialism," says Mikhail. "European ideas about disease begin to take hold in places like Istanbul [and] Cairo." Europeans had strong ideas about where trash belonged—outside the city—and where it did not—on the streets. The French got rid of the garbage in the streets, moving it outside the city. And without garbage in the streets, no one needed dogs to eat it. "They come to be associated with the sources of disease and are thought to be expendable, right, because we want to remove garbage and we want to remove those who participate in the garbage," Mikhail explains.

Not only that, the dogs, which were known to follow the human dwellers of Cairo in packs, saw the French as interlopers and formed their own canine resistance. They barked at French patrols and drove them out of alleyways. The French troops quickly had enough. On November 30, 1798, French patrols took to the streets of Cairo with baskets of poisoned meat. In the morning, canine carcasses littered the streets.

The Ottoman Empire came back to Egypt around 1805. But they returned with a different frame of mind. "The leaders of Egypt . . . grab on to these ideas [that] a civilized city should operate in this way, you know, garbage should be removed from cities, et cetera," Mikhail says. "And so for all these reasons, dogs get cut out of the productive roles in society that they used to play."

Urban, unowned dogs have come to be seen all over the world—regardless of religion—as signs of a city that's in disorder. An unowned dog is perceived as a dangerous thing. It might spread disease. It might bite someone. It must be taken care of.

LIKE THE DOGS of Cairo and the rabbits of Australia, pigeons used to have a job. Rabbits were supposed to run in front of a gun, dogs were trash control, and pigeons were messengers, guano producers, and food. Now, if they do anything, it's in the realm of trash control, picking up our dropped pizza crusts. But in the European sanitation model, that's not something people value anymore. The pizza crust shouldn't be there, but the pigeon shouldn't be there eating it either.

Exclusion, shooting, or poison can protect one building for a while. But if you want to really reduce pigeon numbers, you need to tackle what they eat and how they breed. "It's pointless going to a council and saying 'don't kill birds,' and not having a solution," says Guy Merchant. Merchant is the founder of the Pigeon Control Advisory Service (PiCAS Group) and now runs Bird Control Consultancy Services International (BCCSi) out of Ely in England.

His introduction to pigeon control came from helping out at an animal shelter. After rehab, pigeons tended to stay in the area, taking advantage of the abundant shelter food. The idea Merchant came up with wasn't to make them leave, but to control their numbers. He put a set of nesting boxes high up on a wall. When the pigeons took the housing and began laying eggs, he went in and stole them, using them to add pigeon omelets to the diets of the other shelter occupants. But there was one hitch, which used to be an advantage for farmers cultivating pigeons. They quickly found that if you take away eggs or nestlings from pigeons, the pigeons will make more eggs. This works great if you want squab in your diet. But the pigeon eggs quickly overwhelmed Merchant. His pigeons produced so many eggs, in fact, that the local lady pigeons started to suffer from calcium deficiency, all the calcium in their bones going to the eggs. So Merchant started replacing the eggs with little plastic dummy eggs, to make sure the pigeons didn't make more.

Breeding facilities don't work on their own. If you really want a pigeon problem solved, Merchant says, "it's down to food supply." That's the key to the urban niche, where humans reliably provide rich piles of food. And this, Merchant notes, is where people fail.

"We still haven't understood that people are the problem, not pigeons," he says. "If we control humans, by default we control pigeons, but that's incredibly difficult to do."

Merchant recalled one of the most intense pigeon infestations he'd ever seen, at a fort turned fancy hotel in Jaipur, India. "You've never seen pigeons until you've got to India," he says. At this swank hotel, the pigeon problem got so bad "they had two guys employed all day running around the site with red flags" to scare the birds away, he says. "That was their control mechanism."

Those guys were probably thrilled when Merchant was called in to find a permanent pigeon fix. He quickly realized what he needed to find out first. The pigeons loafed at the hotel. But they didn't lunch there. Where were they getting their food?

He found the answer about one kilometer away, where a local Hindu temple opened its kindly doors every morning for a pigeon breakfast buffet. At dawn, the temple opened huge sacks of corn. The pigeons got a hearty meal, then headed back to the hotel to loaf and breed the day away. Merchant convinced the temple to stop feeding the pigeons, and the pigeon population took care of itself.

Similarly, when Trafalgar Square decided in the 1990s that the pigeons were more of a problem than a tourist trap, the city hired falconers to scare the pigeons away. But that's not what really took the pigeons off the statue of Nelson. The city also took away the permits that allowed vendors around the square to sell birdseed to tourists. Take away the food, take away the pigeons.

When working with city councils and large organizations, Merchant says, solving the problem is often very simple. Some processing plants that leave enough food waste to make a pigeon paradise, however, will never care enough to put in effort. And individual people, individual pigeon lovers, are the most complicated of all.

Lots of people like to feed the birds. Carrie in Harvard Square is only one example. "One elderly gentleman had been coming in every day to a [city] for many years with a shopping trolley," Merchant says. "He was distributing somewhere between forty to fifty kilos [of bird food] every day. One individual can be responsible for sustaining in that situation three thousand to thirty-five hundred pigeons," he explains. "That's a lot of pigeons."

Then, it becomes important to ask, why do people feed pigeons? Some—parents with kids, or a random person on their lunch break with a sandwich—will leave a crust for the birds, Jerolmack writes in his book *The Global Pigeon*. But others, like the man Merchant knew, are "focused" feeders. They are not just offering a crust, they're buying bags of birdseed and carting it from place to place.

Often, Jerolmack notes, this intense feeding starts because someone is lonely. He relates the story of a woman named Anna, who fed the pigeons in New York City. As he got to know her, she told him that "all of her old friends from the neighborhood had died or been 'put in a home.'" Many pigeon feeders were like Anna. A study of pigeon feeders in Madrid and Basel said they were "those not accepted in general." There are exceptions of course; Jerolmack found regular feeders included wealthy people, married men, and others.

Giving feels good, and pigeons are always grateful recipients. Some people, wanting that warm feeling, will keep coming back and feeding more, and more, and more.

People who don't like the feathered flocks that result from all this feeding aren't just seeing pigeons. They're seeing the results of the loneliness and lack of care for others in our own society. Merchant finally convinced his man with the shopping cart to stop feeding the pigeons. He says, "I got him to stop feeding but only because he saw me as pro-pigeon." But maybe it was more. Maybe the man also saw Merchant as a friend.

As I talked with Merchant and Jerolmack, I thought again of Carrie, of the way she smiled at her pigeons. Being unhoused can be cold due to more than the weather. Holding a sign that says

GOT EMPATHY? shows that a lot of people don't. Is it any wonder that you'd want to feed a pigeon? The pigeons know her. They're excited to see her. Feeding an animal—pigeon or squirrel or stray pet—feels like doing good in the world.

Carrie is convinced that the pigeons are more than just grateful for her daily feedings. Her voice grows husky as she talks about a pigeon named Søren Kierkegaard who used to visit her. "I loved that bird," she says. "He used to check on me." One day, she remembers, she was sick. It was winter, the week before Christmas. "I was lying down during the day, which I never do," Carrie says. Kierkegaard was up on the ledge with the other pigeons above the CVS. "He kept coming down and landing on top of me. And I kept saying, 'It's okay Søren, you can go back up there.' And he'd go back up on the ledge and then fifteen minutes later he was on me. He did this all day."

When the short Boston day ended before 6:00 p.m., the birds went off to wherever it was they called home. Not Kierkegaard. "He came back at six thirty at night in the dark, he waited for me," Carrie says. "I was off doing things, I thought they were all in bed so I didn't have to rush back. I come back, he was sitting up there all by himself in the dark." Carrie called him down, snuggled and petted him and fed him a little. She told Kierkegaard she was okay and to go to bed. He went back to the ledge and watched her awhile, "just to make sure that I was really okay." Finally, he went to bed.

But the next night, he did it again. "Every night, he did this," Carrie says. "And then two days before Christmas, the same thing happened, and I told him it was okay to go to bed. He went up to the nest, and I never saw him again."

PIGEONS FIT THEMSELVES into our urban lives—a niche where we formerly made them welcome. We adapted them beautifully for our world, breeding them for tameness, for meat, and for carrying messages. Colonists similarly brought rabbits to new places, for meat and sport. When our world moves on, and animals remain, or get away from us, they steadily devolve in our minds. We can exclude them or shoot them. We can give rabbits diseases and poison the dogs of Cairo. But they remain, at home even though they're unwelcome.

When I see Carrie with her pigeons, I'm reminded of what we were, what some animals used to be to us, and what we've forgotten. The pigeons, it seems, still remember.

PART III

IN THE EYE
OF THE
BEHOLDER

5

A MEMORY OF ELEPHANTS

I don't know what I expected. A few shouts from keepers, maybe? A distant rumble of feet, like far-off thunder? But the orphan elephants at the David Sheldrick Wildlife Trust in Nairobi, Kenya, don't need any fanfare. Their feet are silent, and they make almost no noise as they trundle through the brush, trotting in a ragged line. Seemingly oblivious to the thrilled crowd and the rattle of camera shutters, the (relatively) tiny elephants know exactly where they're going. They rush excitedly over to their keepers, who are standing ready with large bottles of baby formula, spiked with a few other things good for elephants, like porridge and coconut. The elephants curl up their trunks, the bottles go in, and the formula disappears so rapidly I feel like somebody should be shouting "Chug, chug, chug!" Each gets one bottle, sometimes two, every three hours around the clock for the first two years of their lives.

After feeding, the small crowd of orphans—ranging in age from a few weeks to four years old, and in size from just above their keepers' waists to above their shoulders—relaxes around a mudhole and several large drums of water while one of the keepers gives a short lecture, introducing the animals by name. It's astonishingly like watching toddlers at day care. One blows bubbles with its trunk in the pool of muddy water. Another gives itself a dust bath, and a third playfully knocks over one of the water drums, just to see the rest of the water gush out. Several pick up and put down branches, clumsily chewing off leaves, or just carrying the sticks in their mouths, because it's always kind of fun to pick up and carry a stick. A dispute breaks out over another water drum, and the wrongdoer is quickly banished by another orphan—the young matriarch of the herd—to sulk for a while on the other side of the pool. There is a lot of peeing and pooping.

It is incredibly adorable. But the goal is for them not to stay that way. These elephants, some of whom were abandoned by their herds, and others who lost their mothers through poaching or human-wildlife conflict, get shown off to their adoring fans in the morning (for a fifteen-dollar donation), and again in the afternoon for those who pay an additional fifty-dollar adoption fee (metaphorical, of course; only elephant photos end up going home). In between, they roam the nearby Nairobi National Park with their keepers, learning how to be elephants.

These are African savanna elephants—*Loxodonta africana*; in Kiswahili, *ndovu*. They seem to live up to all the expectations people from the Global North (countries sometimes called Western or First World) have about these creatures. They are sweet, peaceful, and intelligent.

But elephants are also something else. Two days later, I drive with my friend Shannon and my guide, Simon Mwanza, through the village of Mwambiti. Our host, Derick Wanjala, is a beehive fence officer (more on that later) at Save the Elephants—a charity devoted to, well, saving the elephants. Wanjala stops the car to exchange a few words with two women walking along the side of the road. As we continue on our way, he gestures back toward one of them. "The elephants came and they destroyed her house," he says.

They came in the night, drawn by the sweet scent of fresh corn. The woman had already harvested her fields and was storing her maize inside her house for safekeeping. For hungry elephants, house walls are no obstacle. They crashed right through the wall and destroyed her harvest and her home in one go.

She, at least, escaped with her life. One woman in the village has already died this year from an encounter with an elephant. The villagers suspect she went outside in the night, trying to defend her crops from the invader's appetite. Another ended up with an injured leg. As everyone told me during my travels in Kenya, if you see a wild elephant and you're on foot and undefended? Don't ever get close enough for a selfie. Run.

For those who don't live with elephants, it's easy to think that the only human-elephant conflict there could be is the kind that humans perpetrate, the kind that poaches these beautiful creatures for their oversized incisors. But elephants are also living tanks, capable of killing, disemboweling, knocking down houses, and eating a farmer's entire crop for the season. Human-elephant conflict can go both ways. And in Kenya, India, and other countries, now it's often humans trying to keep the elephants at bay.

Outsiders' beliefs influence how those people go about saving their livelihoods—and even their own lives. When Europeans first colonized the African continent, their beliefs about elephants were as inspiring, sometimes terrifying game animals. They slaughtered them by the millions. As elephant populations dwindled and people in the Global North learned more about extinction, elephant intelligence, and empathy, views changed. The vestiges of the old views live on in continued poaching. The new view of elephants wants to see more elephants—and wants to see them living in the wild, free of human interference.

But elephants don't live in landscapes free of humans. They never have. Now, because people who don't live with elephants have a romantic vision of elephants, the way Kenyans approach conflict with elephants is almost never deadly. It's not just because people living here in Kenya don't want to kill elephants. It's also because the downsides to killing an elephant are so high, both to the people themselves and to Kenya's reputation and economy.

Our beliefs about animals have always affected how we treat them. Sometimes, it's using a live trap in your house for a mouse or campaigning for wolves in the American West. Sometimes it's voting in favor of culling deer or the "removal" of a coyote in the neighborhood. But our beliefs affect the lives of animals—and the people they live with—far beyond our own neighborhoods and countries. And in the case of elephants, the way Kenyans deal with their harms is dictated—with money, tourism, and politics—by the beliefs of people half a world away.

"THEY USUALLY COME at night, and you can hear farmers yelling, shouting, banging the roof," trying to scare them away, says Victor Ndombi, a food-security and conservation livelihood officer with Save the Elephants. Elephants, of course, are well acquainted with these scare tactics. "When you go in the morning you can see the whole land, it's just flat. Just flat." A small herd of elephants can take down an entire crop of maize in a single night—and with it, a farmer's income and the food needed to feed their family.

Conflicts like these are increasing in Kenya and well beyond. Both Africa and South Asia—which has the Asian elephant (*Elephas maximus*)—have growing human populations that overlap a lot with their native wildlife. In India, where populations of elephants live alongside high concentrations of humans, both people and elephants suffer. Many of the humans killed by elephants are the people nearby when elephants come out of jungles to eat crops—a behavior a human might see as a raid, and an elephant might see as an extremely efficient foraging strategy. If a human is in the way, the human isn't going to win that fight. Between 2014 and 2019, 2,398 people died in India—trampled and torn apart by elephants. Nearly 450 elephants died, too, They are shot in revenge with guns, or electrocuted when scared and grieving people set their electric fences to kill.

In Kenya, elephants raid crops. But they also come into conflict with herding communities such as the Samburu and Maasai when they compete for water and food. They can even cause stress with their very presence. When elephants are wandering the area, you don't necessarily want your kids to walk to school. Local elephants trap people in their homes fairly often, says Danson Kaelo. He conducts interviews on people's attitudes toward elephants in the Greater Mara Ecosystem—a massive area in southwestern Kenya that contains Masai Mara National Reserve and continues south of the border into Serengeti National Park in Tanzania. The day before we spoke, he says, Kaelo had been about to meet a Maasai elder. "And he told me that he cannot get out of his home because elephants have surrounded the boma," he explains, referring to the enclosures people make around their family compounds. When a black bear traps a family in their home for forty-five minutes in the United States, it's news. When an elephant herd in Kenya traps a family and their cattle in their home and corrals for a day, it's Tuesday.

Elephants may threaten, but they are also under threat— partially due to poaching, and also due to being hemmed into small areas with human populations. Both the Asian elephant and African savanna elephant are endangered, and the African forest elephant is critically so. There are only about four hundred thousand African elephants of both species left. In a 2021 map, Enrico Di

Minin and his colleagues showed that 82 percent of conservation areas with lions and elephants in them are next to areas with lots of humans, lots of cattle, or lots of agriculture. Countries where the risk of conflict is highest—with humans, cattle, and agriculture together—also play host to 66 percent of the remaining elephants. In their map showing conflict risk, the "severe" ratings form exact outlines of many of the national parks and reserves in East Africa.

"Many people are thinking about how to protect elephants, but not how to protect people," says Wilson Sairowua, a conservation officer at the Mara Elephant Project in the Lemek Conservancy in Kenya. Efforts to protect elephants from poaching in the Mara ecosystem, he says, have been incredibly successful. But as poaching goes down, conflict with humans goes up. "Conflict is bigger now," he says. "We are actually forgetting about poaching."

There are plenty of people who benefit economically from elephants without killing them, of course. Tourism makes up 10 percent of Kenya's economy and employs about one million people. And most of the people coming to Kenya from all over the world are coming to see the wildlife. They especially want to see the "big five"—lions, buffaloes, rhinos, leopards, and of course, elephants. "Down here in the Masai Mara the biggest investment is in tourism," explains Jonah Noosaron, a member of the Maasai and a pastor in the town of Talek.

People living with elephants do not hate elephants. They recognize that the animals play an important role in the ecosystem. They also recognize their beauty. Elephants are part of their culture, stories, and traditions. But their beliefs about elephants are not the starry-eyed stories of peaceful wisdom or playful behaviors we have in the West. It's hard to remain starry-eyed when death and destruction stares you in the face. Purity Taek runs surveys around Masai Mara for scientists studying human-elephant conflict. She says Western tourists surprise her with their views on elephants. "Some of them think that elephants are not that bad," she says.

They think "elephants are friendly to the environment, friendly to the people around." Taek respects elephants and the role they play in Kenya's wildlife and economy, but "friendly" is not a word she would use. And sometimes, frustrated, she wonders, "What type of elephants do they have?"

WOLVES, LIKE ELEPHANTS, are animal celebrities. Wolves have groupies in urban areas, eager to see a tiny bit of wild restored to the American West. Elephants have people in the Global North who love the idea of peaceful giants roaming a human-free landscape. Both species bear the weight of human beliefs—both about the animals themselves and about what "wild" animals and "wilderness" should be like.

European colonizers travel with a lot of baggage. Some of that baggage is in the form of beliefs about the animals they encounter. By 1621, when men from the Plymouth Colony came upon wolves, the canines had been hunted out of England for at least a generation. Like modern people whose closest wolf encounter is watching Animal Planet, the colonists only had stories to go on. These were not lovely stories of wolves as the epitome of wildness. Instead, as Jon T. Coleman writes in *Vicious: Wolves and Men in America*, many of the legends surrounding wolves were about horrible, hairy criminals. Wolves blew down the houses of defenseless pigs and showed off their teeth to girls in red hoods. The Bible taught colonists that they and their livestock were sheep, helpless prey against the wolfish sinner.

In 1814, John James Audubon (yes, that Audubon) saw a farmer trap, hamstring, and throw three live wolves to his dogs. Neither Audubon nor the farmer felt one iota of guilt, Coleman writes. They "shared a conviction that wolves not only deserved death but deserved to be punished for living."

This was completely at odds with the way local Indigenous

groups dealt with wolves. Some, such as the Narragansett, did kill wolves, Coleman writes, but not indiscriminately. Instead, wolves were killed when they stole deer that people had snared. A life for a life. "The Indians destroyed individual wrongdoers," he writes. "The English punished an entire species."

Wolves are dangerous, of course. There's a myth that no human has been killed by wolves in the documented (ahem, colonized) history of North America. It's laughably untrue. Attacks are rare (much rarer than brown bear attacks, which take at least one or two lives per year), but they do happen. The most recent reported death was in 2010, in a town called Chignik in Alaska. A schoolteacher took off on a solitary run. She never came back.

More often, of course, wolves eat livestock. In 2015, a USDA report shows that 41,680 cattle fell to the teeth and claws of predators. Of those, 4.9 percent, 2,040 cattle, were killed by wolves. It might seem like a lot. But in 2015, there were 112 million head of cattle in the United States. About 3.9 million of those were lost that year, mostly to disease or other health problems. The 2,040 dying to wolfish jaws is only 0.05 percent of all the ones who died before they could become burger.

But numbers aren't experience. Each loss is a sad, often very young, animal lying on the bloodied ground. Profit margins in agriculture are slim, and people who go into the work do it because they love the outdoors, and the animals they care for. Each loss is an emotional one, in addition to the loss of money that people need to survive.

Some urban and suburban dwellers find those emotions easy to overlook. They are emotions felt by people not like them. People with different priorities and ways of living. When people don't live near wolves, "it gets to be just sort of the Disney version, you know, that wonderful animal out there that's symbolic," says John Shivik, a predator biologist now working for the U.S. Forest Service in Logan, Utah. "And people will tell you that just knowing that those animals exist is a value. I get that, but knowing that the animal exists, you're going to have to acknowledge that that thing exists *somewhere*. It's a tangible reality and impacts people too."

In his book *The Predator Paradox: Ending the War with Wolves, Bears, Cougars, and Coyotes*, Shivik tells the story of a farmer in Wisconsin who has lost two cows to wolves. She's a widow, he tells me, who only had about six cows total. "That was how she's going to eat the next year," he says. "A lot of producers aren't just the big rich cattle barons that don't care one way or another." It's not just monetary loss, either. It's "a real calling for them to do the work, and a way to raise their families outdoors," he says.

Those people can end up backed by lobbying groups that demand wolf hunts and bounties. And here, notes Adrian Treves of the Carnivore Coexistence Lab, the human-wolf conflict becomes human-human conflict. The people who live with wolf-induced losses and wolf-induced worries are angry at the liberal elites in cities who want to watch wolves in documentaries and the Indigenous peoples who see wolves as a necessary and important part of their ecosystem and culture. But the groups facing off are often faceless. There's no one person specifically to yell at, to try to convince. And so ranchers living with wolves vent their frustration on the animals close at hand, the animals that are causing damage just by being themselves. Game managers call this technique of predator management "shoot, shovel, and shut up."

Some of the discrepancy in treatment might be the difference between a carnivore and an herbivore, says Jenny A. Glikman, who studies human-wildlife interactions at the Instituto de Estudios Sociales Avanzados in Spain. "There is definitely a difference in terms of tolerance." Vegetarians are supposed to be gentle, meek, and even weak. Meat eaters are strong, bloodthirsty, and violent. Maybe the idea of humans being a wolf's prey, she says, makes us that much less willing to put up with wolves being wolves near us.

But it could also be that people in the developed, industrialized, walled-off-from-nature Global North have a lot less tolerance for animals that harm them and a lot more tolerance for animals living far away, harming other people, notes Susanne Vogel, a conservation scientist at Aarhus University in Denmark. If elephants lived on the Great Plains of the United States and tried to eat our amber waves of grain? "They'd be shot before they could start," she says.

If people in Kenya thought about elephants the way some ranchers in the United States thought about wolves—as irredeemable predators—those losses could very well mean elephant slaughter. Compare American livestock losses to the total loss in livelihood that people experience in Kenya after an elephant raid. A house flattened. A farmer dead. And a history behind it all, with disputes over land rights and damage compensation.

Neither the American West nor the savannas of Kenya are anything close to human-free, and they haven't been for thousands of years. If people are pursuing a hunting-and-gathering lifestyle, wolves can teach them a great deal, as Joseph Marshall III notes in his book On Behalf of the Wolf and the First Peoples. "The first peoples understood that while they could emulate the wolf and be like the wolf in some or many ways, they would never actually occupy the place the wolf had," he explains. "Furthermore, they understood that they had a power to understand and that this capacity set them apart from other species."

Marshall, a Lakota elder from the Sicangu tribe, notes that of course conflict happened, but wolves as a species were not a problem. "We coexisted. That did not necessarily mean that we were always pleased with one another. But it did mean that we always respected one another's right to be."

This coexistence was part of a society in which most Indigenous groups didn't keep herds. But they had many layers of coexistence with animals—including the bison many tribes relied on for their livelihoods. Europeans brought pastoralism—herding sheep, cows, and more—with them, and with the practice, a different perspective. They made their living off herds of wolf prey. That prey and the need to protect it came with an entirely different view of their fellow predators.

Human pastoralists previously coexisted with herbivorous elephants in Kenya, says Martin Maina, a member of the Maasai and a political science consultant in Lawrence, Kansas. In fact, he notes, they had a symbiotic relationship. The Maasai could make use of elephant migration paths, for example. "Fifty to one hundred of them would use the same track over and over and over and over and eventually it forms a channel which is almost like a stream," Maina explains. When the rainy season comes, water funnels down those tracks, and so the Maasai set up their bomas near them to take advantage of the water source.

Elephants also pruned the acacia trees. "If they are not maintained, really well pruned, they are really susceptible to start savanna fires when struck by lightning," Maina says. By keeping the trees in check, elephants kept fires in check too.

It wasn't all peaceful coexistence. When a male elephant is old enough to breed, he enters a state called musth periodically. If you thought human adolescence was bad (acne and hair where there was no hair before), be grateful you're not an elephant. Oily secretions run down the sides of their faces, fluid drips from their penises constantly, and their testosterone levels spike up to sixty times higher on average than normal. Elephants in musth are—perhaps understandably—very irritable. Scientists, in fact, know very little about musth—because no one wants to get close enough to find out.

During this aggressive period, Maina explains, male elephants are more likely to cause trouble for people. "They charge at anything and everything," he says. "So if they come across a Maasai herd of cows, they will kill cows." In the past, before poaching had severely thinned elephant populations, the older, bigger males would control the temperamental youngsters—partly because older males prevented the young elephants from going into musth as often.

Then came the poachers. "The poachers would target the elephant with the biggest tusks. And that will usually be the matriarch and the elephant bulls." Without the big, elderly males to hold the younger ones in check, Maina says, relationships were bound to get worse.

Elephants also began to have an indirect negative effect on the Maasai, Maina notes, because of the way they had been initially conserved. When British colonists started Kenya's national park system in 1946, there was no place for Indigenous people in it. "The idea was to remove local Indigenous people from the environment," he says, to make the park "as nature intended it. I don't know what that even means." The Maasai were not allowed to graze animals on their traditional lands—lands that had originally been shared with elephants, but were now the elephants' alone.

The government at one point even tried to restrict how many cattle Maasai could own, limiting families to one cow per acre, and 150 at most, explains Pastor Jonah Noosaron. That didn't go over well. To the Maasai, cows are measures of wealth. "You will not wish to reduce your cattle," Purity Taek explains. "Some people have fifteen hundred cows."

It's easy to see why resentment toward the protected wildlife might grow. The Maasai "wish they could be free, just grazing everywhere they used to," says Pastor Jonah Noosaron. The pastor is Purity Taek's husband. He and Purity are both Maasai, and Noosaron of course owns cows.

But Noosaron is now part of a human-elephant coexistence plan. He leases his land to the Mara North Conservancy. Kenya gained independence in 1963 and rewrote their constitution in 2010. In the process, the Maasai regained governance of their traditional lands—including the nature reserves. But they don't graze their herds wherever they like. Instead, the Maasai have continued to protect the reserves—and to pool the lands they individually own surrounding the reserves into conservancies, providing even more protected land for wildlife. They may not graze where and when they want, but "they're benefiting from it," Noosaron says. The pastoralists profit off the lease, and retain the right to graze their cows in parts of the conservancy. In that profit, they also materially benefit from the tourism around elephants.

There are criticisms, of course. Some conservationists worry that the human activity in conservancies squeezes wildlife into the center of the protected areas and harms the function of the ecosystem. Others note that as more and more land is set aside to form conservancies, many Maasai communities end up "gentrified," and their ways of life stigmatized, as conservationists claim their cattle eat all the available grass and leave the elephants in crisis. In next-door Tanzania, Maasai herders are still getting forcefully evicted in the name of wildlife. It's far from a perfect system.

FOLLOWING KENYA'S LONG history as a British colony and its subsequent industrialization, the silent agreements between humans and elephants haven't been restored. Instead, with the Western beliefs about elephants funding much of the research, new forms of coexistence are coming to be.

Coexistence can't always mean that humans and elephants are near neighbors, notes Lisa Naughton, who studies social conflict and land use around protected areas at the University of Wisconsin–Madison. She works with "a multitude of people whose take-home message of every paper is, 'we shouldn't think of ourselves as separate from nature,'" she says. "And I think that's a well-intentioned remark . . . but I think it can also be sort of unrealistic." It's not wrong. We are not separate from nature. But not all animals go well with other animals. Elephants—or wolves—and the ways people currently inhabit landscapes don't always go together. If humans won't change, something has to give. "There are some animals that are really hard to live with," she says. "Coexistence means a little bit of separation."

Electric fences are one way to get that separation. Run enough volts through a wire, and it'll give even an elephant a truly unpleasant zap. Electric fences are going up all over Africa and Southeast Asia to try to stop human-elephant conflict.

They do work. The fences are tall, spiked with barbed wire, and most other animals avoid them instinctively. But they're also hugely expensive to build and maintain. They run the risk of cutting off elephant travel routes. And elephants will still go through them, if the motivation is high enough. They use their nonconducting tusks to short-circuit the wires. Some people, faced with raiding elephants, even modify the fences to kill. So Lucy King, now the Elephants and Bees Project leader at Save the Elephants, wanted to try something a little more natural.

There is this oddly persistent myth (thank the Roman historian Pliny the Elder) that elephants are afraid of mice. They aren't. They are, however, definitely not fans of bees. With skin that's about two centimeters thick, you'd think elephants would have nothing to fear. But the thin skin around their eyes and the sensitive tips of their trunks are another matter entirely.

King and her mentors, Iain Douglas-Hamilton and Fritz Vollrath, learned from Maasai living in Laikipia that elephants avoided feeding on acacia trees that had bees in them. King then played recordings of African honeybees buzzing to herds of wild elephants. She found that 94 percent of the groups vacated the area. Furthermore, almost half of the elephants didn't just saunter off—they ran.

Imagine what a whole pile of beehives could do. King began to test borders of beehives around people's farms. In the first pilot study, a fence full of beehives reduced successful elephant raids by

86 percent. Beehive fences have gone up in twenty countries and counting.

The original beehive fences were natural tree trunks, cut, hung on their sides, and full of bees. By the time I get to see the ones at Samuel Kala's farm in Mwambiti, the program has gotten a glow up. The large patch of dirt (it's the dry season and the crops have been harvested already) is surrounded by a string of bright yellow boxes. Half are beehives. The other half are dummy hives—pieces of yellow-painted plywood that take advantage of elephants' relatively poor eyesight. "When [the elephants] see a hive and the dummy, they just assume they are the same thing," Derick Wanjala, the beehive fence officer, explains. "So they try to avoid everything."

The fence shows just how much King and her colleagues have been learning. Each hive is partially shaded by spindly trees to keep it cool. The top is protected by a stout wire cage to protect against hungry honey badgers. All of this is to entice wild bees to move in.

The beehives are connected by a wire strung up around the field. Elephants come by and "knock," Wanjala says. They'll nudge the wires a few times. If no bees are home? Elephant dinner is served. Wanjala wiggles the wire, vibrating hives down the line. Curious bees start coming out of the nearest hive. He hastily stops shaking the wire.

The Elephants and Bees Project provides the hives, the wire, yellow paint, and a bee suit. It's up to the farmer then to make the dummy hives, cut or buy the poles to support the hives, keep the hives clean—and attract the bees.

Kala is one of the best, Wanjala says. Ten of his sixteen hives are occupied. Kala, striding over from a meeting, says he hopes to harvest the honey from all his hives this season. Per season (there are usually two per year), he'll get about eight thousand Kenyan shillings from his hives—about $80 at the time of my visit. That's pretty good money, in an area where many people get by on around $2,500 per year (in 2009 dollars, which in 2021 is about $3,150 per year). It's such good money, Wanjala says, that some farmers have just gone all the way over into beekeeping and stopped growing maize.

The elephants are a constant, Kala says in Kiswahili as Mwanza and Wanjala translate. The pachyderms don't wait for full-sized ears of corn, they'll come eat even the young plants. "And still the elephants will follow your harvest in the house or the granary," Mwanza adds. People often don't get a full night's sleep for three months straight, having to chase away the elephants. It's a potent reason to start a beehive fence—and to shift away from corn growing entirely.

Elephants and Bees only has so much funding, and the bee-fence setup is expensive. Bees can't be depended on to move in. And some resident elephants are getting wise to empty hives. So Wanjala has been looking into other options. He's trying a small white-noise machine that will play the sound of bees when elephants hit the wires, so the bees themselves aren't required. And he's trying another deterrent—an elephant repellant.

Wanjala recently hosted a group of farmers from Uganda to lead a class, teaching local farmers to brew a potent mix of smells designed to keep elephants at bay. Here's your recipe: Boil some water and add chilies, ginger, garlic, and neem-oil. Steep for a while, then mix with some elephant and cow dung. Mix that with some rotten eggs. Add cooking oil and let the whole thing ferment in an oil drum set out in the hot equatorial sun for a couple of weeks.

When I ask to smell it, Wanjala and Ndombi look at me with grave misgivings. I insist. They show me some other things, hoping I will forget. I do not forget. Finally, Wanjala and Ndombi reluctantly lead me over to a giant blue plastic drum. They unscrew the lid and stir the contents with a stick. I lean in.

At first, it's almost tasty. Then, suddenly, it's *really* not. The notes of garlic, chili, and ginger give way to fermented old eggs and poo with a speed that is eye-watering. It is, indeed, repellant.

This potent mix gets poured into plastic water bottles which have holes near the top. The bottles are strung up on a wire fastened to the fence around a just-ripening crop. When elephants come up and jostle the fence, the liquid inside the bottles sloshes, and the smell comes out. The elephants, with their highly superior sense of smell, gag and leave. In Wanjala's current trials with eight farmers using this method in Mwambiti, all managed a full harvest. It's a good bit cheaper than a beehive fence. The Elephants and Bees Project provides the more expensive ingredients like ginger, and fence posts are easy to come by. Plastic bottles—which litter the roadside in profusion—are even easier.

Another option to prevent elephants from crop raiding is to just not have crops to raid, as the farmers who have abandoned corn for bees have shown. Esther Serem is a Women's Enterprise project officer at Save the Elephants. Under her guidance, women in the surrounding villages are banding together in groups to do microbanking and are diversifying their skills with courses in sewing bags and making baskets. She is also working on family planning, so there's less pressure to find the school fees to send so many kids to school.

Victor Ndombi teaches children and adults how to grow vegetable gardens and raise poultry to sell in the community, instead of growing crops like maize for subsistence. And the whole team encourages villagers to grow crops that are not elephant candy—crops like chilies or sunflowers that elephants don't like.

The whole idea, Serem says, is to make sure the villagers aren't relying on elephant-attracting crops as their only source of income. Take a woman named Eunice, she says. She was a subsistence farmer who discovered she had a talent for basketmaking. "Now that she's moving to the baskets, she's growing a smaller portion of the farm," Serem explains. She doesn't have to grow much, only a "little maize for just roasting and a small amount of beans to be able to feed her family." Those can be grown close to the house—a

less tempting prospect for elephants. She might even shift to growing chilies or sunflowers and selling those for money to buy food.
That would mean she could sleep through the night in peace.

The Elephants and Bees Project is a wonderful example of
how efforts to keep elephants away from human crops can benefit
both humans and elephants. But it also shows how many hurdles
need to be overcome to make it happen. Someone has to pay for
the hives. Someone has to help farmers harvest the honey. The
project leaders soon realized that the farmers were not able to sell
honey themselves at a profit, so Save the Elephants also has to
process the honey and sell it (often to Western visitors who come
to view the hives, but also to locals who consider honey as essential dawa, or medicine). The alternative crops—hot peppers and
sunflowers—require markets; so Save the Elephants had to create
those markets with sunflower oil producers. Serem's and Ndombi's programs teach local women how to earn alternative income
streams, but those gardens take money to start, and the bags and
baskets require training and people who will buy them (again,
Western visitors). The result is that Save the Elephants has had to
catalyze its own microeconomy, all to keep humans and elephants
apart.

Beehive fences won't work everywhere. Not every farmer wants
to raise bees or make baskets or pursue alternative crops. And the
elephants have it right. People are also often afraid of bees. Even
if someone isn't afraid of bees, beehives aren't necessarily a good
idea if you have a family with a lot of kids who could get stung, says
Wilson Sairowua at the Mara Elephant Project.

Sairowua works in the Greater Mara Ecosystem in the Lemek
Conservancy, next to the Masai Mara National Reserve. It's the
iconic savanna that many Western people think of when they think
about where elephants live. Driving through it feels like entering a
Dr. Seuss book, with horned and betusked animals grazing in long
grass, and tall, flat-topped trees rising up at intervals.

Farmers around the edges here also grow maize for subsistence, though Sairowua notes that some are switching to sugarcane. If you thought corn was an elephant draw, that's nothing to sugarcane. When elephants get into the sugarcane down by the river, Sairowua says, "You can even hear elephants up here making a funny sound. They're very happy, enjoying the sugarcane." The farmers, however, are not remotely pleased.

Local farmers also use smelly fences, Sairowua says. Theirs are bottles with a mix of chili and engine oil. But eventually the elephants realize it doesn't mean much of anything. "Someone told me that elephants will come in and all just stand next to the fence," he says. "And then they will push the babies through. Once the babies are in the mothers will go." Not only are they shoving their babies in the direction of delicious maize or sugarcane, they are also teaching them that chili pepper fences aren't worth fearing. Hold your trunks, kids, there's food on the other side.

That's when groups like the Mara Elephant Project take over. A now independent offshoot of Save the Elephants, it has a special kind of fence—a geo-fence.

It's not a fence people or elephants can see. Instead, it's a fence made of collared elephants—or rather, the information they transmit. Right now, the Mara Elephant Project tracks more than fifty "flagship" elephants. These are elephants that live in high-conflict areas, like on the borders of the conservancies and parks. Some elephants in large herds also get collared to provide information on the herd's location. And of course, if any particular elephants are especially into chomping on corn and sugarcane, they're highly likely to end up with a collar.

At the Elephants and Bees Project, I got to heft one of their collars (they also run a collaring program). A large box in the middle holds the GPS tracker, attached to a thick strip of leather like the belt on a weight-lifting champ. When applied to an elephant, a counterweight is mounted on the bottom, so the GPS tracker always points toward the sky. It's easily thirty pounds. But to an elephant, it's probably as light as my wristwatch is to me.

As Sairowua talks, he glances repeatedly at the large Apple Watch on his own wrist. It's vibrating almost constantly with Whats-App messages, each a notice of a collared elephant on the move. "Every time an elephant crosses the geo-fence, it sends me a message on WhatsApp," he says. Should one get within a kilometer or so of a farm or community, a set of local rangers springs into action. They drive out to the straying elephants and start trying to force them away. They play drums and set off firecrackers, using the noise, lights, and smoke to send them back toward the parks and conservancies. If those don't work, "we also have the chili balls," he says. These are little nuggets about the size of Ping-Pong balls, made of chili and dry charcoal. When thrown at an elephant, they burst.

If the chili balls fail, the Mara Elephant Project deploys three quadcopter drones. "The elephants are scared of the drones," Sairowua says. "Because of the sound, it's like a bee sound. . . . So whenever the elephants hear the drone they run."

Elephants are notoriously smart, though, and are rapidly learning that drones don't sting; some are even trying to pull the drones down. So if all those tactics fail, it's time for the helicopter. Marc Goss, the CEO of the Mara Elephant Project, is also its helicopter pilot. Flying low, elephants are faced with the incredible noise and the hard downdraft of the propeller blades. This always works. Unfortunately, it's also expensive. The elephants don't always go without protest either. Some have started lifting their trunks toward the choppers' landing gear. If they ever grasp it, they could easily bring down the helicopter, with disastrous consequences for the pilot.

Which is why the Mara Elephant Project is looking into other methods. The current effort is a drone carrying chili spray. If the elephants don't listen to the drone or try to pull it down, a ranger can hit a remote control, and the drone will spray chili right up the questing trunk. (If you learn one thing from this chapter it's this: If elephants learn to like spicy food, their world takeover is at hand.)

Other groups try similar tactics, hazing elephants away with noise, chili balls, and Range Rovers, with the addition of things like rubber bullets, flaming torches, and more. Local farmers will also chase elephants with spears and arrows, trying to injure the truly determined pachyderms enough to make them leave. But there is one thing they will almost never do. They will not shoot to kill.

KENYANS FACING HUMAN-ELEPHANT conflict are fully aware that the elephants are valuable. In fact, they are often too aware.

There's a certain amount of resentment, explains Danson Kaelo. If an elephant gets killed when people try to chase it off, he says, the killer faces a fine of up to twenty million Kenyan shillings, or two hundred thousand dollars in 2021, and life in prison. If the Kenya Wildlife Service catches poachers toting guns in a protected area, they can shoot first and ask questions later. Anyone caught trading ivory is subject to life in prison.

If an elephant kills a human, their family is also entitled to compensation. They can apply to the government, and with a lot of paperwork and luck, get about five million Kenyan shillings, or fifty thousand dollars in 2021, Kaelo says. For a farmer, that's a lot of money. But it's not easy to get. "Following up for that money is very hard, it's very hard," Kaelo explains. "For some families it can take years."

Journalists love to say that "if it bleeds, it leads," but in Kenya, a bleeding elephant also leads a lot more than a bleeding person. In Tsavo, the "community really felt marginalized," says Lydia Tiller, a research and science manager for the Elephants and Bees Project at Save the Elephants. "You know, if an elephant was poached, there'd be choppers and thirty people on the scene. But if someone got injured by an elephant, there'd be no one there."

It can breed frustration—and anger. The elephants are technically the property of the government. In fact, people resentful of elephants in places such as Namibia sometimes refer to them as "the government's cattle." In June 2022, farmers from the Kenyan town of Mashuru—a village north of Amboseli National Park—blocked the Nairobi-Mombasa highway, the main artery from the capital to the coast, with burning tires and rocks. They were protesting human-wildlife conflict—asking for answers to the elephants eating their crops. The county governor asked the Kenyan wildlife authorities to "keep their animals away from our people." Security forces shot and killed four of the protesters.

It's easy to see why people living with elephants feel resentment. It's hard to escape the conclusion that, in Kenya, elephant lives are worth far more than those of the Kenyans who live with them. Especially because, in terms of cold, hard cash—twenty million Kenyan shillings for an elephant, five million for a person—they are.

The fine for killing an elephant is so high because it is so valuable—to biodiversity and to Kenyans, but also to the people from the Global North who want to come and see the animals they consider so lovely and peaceful. To them, elephants are never pests. Elephants are never in their backyards, and the havoc they cause is a price the tourists will never pay.

The piles of donations to save the elephant, to protect it from the people who would kill them for their own livelihoods, are a vestige of a colonial system that says that people suffering from elephant conflict in Kenya are not worth nearly as much as the elephants they live with. Some of the NGOs and nonprofits that swoop in from outside the country also send another message—that Kenyans do not know the value of their own wildlife, that they need to be educated out of their backward views and cannot possibly handle their wildlife on their own.

"Funding comes from the West, and it's driven by donors,"

explains Tiller. "That's why I think a lot of the time a lot of conservation programs aren't successful, because it's what people in the West think should be happening, not actually 'What does the community think?'" We in the United States would not want someone from Kenya coming in and telling us what we need to do with our bison. It's no surprise Kenyans feel the same way—and feel that their local context and experience is getting discounted.

Still, there are lots of benefits to the donations that come in to places like Save the Elephants. At Samuel Kala's farm in Mwambiti, one of his beehives has faded Sharpie on the side that says "Funded by Disney." "Most of the donations come from abroad," Kala says in Kiswahili, as Wanjala translates. "If not for you people, it wouldn't be here." Before the beehive fence, Kala explains, during the harvesting season, he never slept through the night due to the elephants.

Being able to coexist with elephants successfully can change people's views, Kala notes, Mwanza taking over translation. Before the bee fence, Kala didn't have an especially good opinion of elephants. "But today with a fence [Kala] doesn't see the damage and now he can respect our wildlife," Mwanza translates. Kala sees the benefits they have to the economy via tourism, and experiences those benefits directly, through selling the honey from his hives.

As Kenya has invested in antipoaching efforts and encouraged the elephant populations to grow, Mwanza explains later, it also had to invest in training the societies around those elephants. Kenyans teach other Kenyans not just how they can live with elephants, but why, and that damage will be compensated—even if that compensation is hard to get and painfully slow to arrive.

Purity Taek has been organizing and running surveys to understand what people really think about elephants, for scientists such as Susanne Vogel. Taek translates the survey questions into Maa, the local language, trains surveyors such as Danson Kaelo to interview subjects, and conducts interviews herself. She and Vogel have found that people living near elephants do not see them as unalloyed villains. They acknowledge the problems of living with elephants, but they also see their benefits. Statements from people interviewed include things like "Elephants . . . should be preserved because they are very important to our environment. They are also very important because our children are working in the tourism industry in the Mara because of them."

Other Indigenous groups, such as the Samburu in northern Kenya, are practicing elephant conservation within their own communities. The Reteti Elephant Sanctuary is run by the Namunyak Wildlife Conservancy, and the elephant orphans are raised by members of the community. The conservation benefits elephants, provides local jobs—and of course tourists can't resist it, though like the other conservation work in Kenya, a lot of its funding comes from donations and organizations in the Global North.

Elephants also have cultural value, Taek and Noosaron note. For people who still believe in the Indigenous religions, finding an elephant placenta from a new birth and burying it in the boma is exceptionally good luck. They consider elephants to be incredibly smart and similar to humans. Elephant dung, skin, liver, milk, and other parts are used in traditional medicine. When an elephant dies, members of the Maasai sometimes cover the body with twigs and pray over it to receive blessings.

It's a complex situation, and there's no one answer that makes everyone feel good. Early colonizing brought in elephant hunting. It introduced economic systems that began to strip the elephant of its habitat and Indigenous people of their livelihoods. Now, because of modern Western beliefs of elephants as loving and brilliant, people in the Global North have spent millions of dollars to help bring these beautiful beings from the edge of extinction. Luxury tourism to see elephants forms a large part of the Kenyan economy. Those same groups also help to fund research and strategies to keep these elephants away from people, to make sure they do less harm. Because of their beliefs, the ways people deal with elephant conflict are much more peaceful than they might be otherwise. Conservationists might be engaged with a technological arms race against a leathery, sentient tank—but they are better and better at reducing conflict, and the methods are better than dead elephants.

However, because of other beliefs—often false beliefs that the people who live with elephants don't respect them, that they don't know how to manage their own ecosystems and wildlife—Western ways of helping can be paternalistic and are built on a history of colonialism.

People in Kenya respect elephants very much. The animals romp across their fabrics and thunder through their cultural iconography. Those living with them also know that elephants aren't always gentle giants. But what we believe in the Global North means that an elephant—to us—is not a pest. And for those living with them, how they coexist comes down to who pays—and whose beliefs are most valuable.

6

A NUISANCE
OF CATS

A few years ago, a stray tabby cat picked her way gingerly through my backyard. She was shabby and thin, jumpy and timid. By the time I'd opened my back door to try to see her more closely, she'd already fled.

I'm a hopeless cat lover. I left out some food. The next day, the tabby was back, eating my offering with tension quivering in every hair. I peeked out at her through my back window. She was a charming molasses-cookie brown under her stripes. Her left ear was cut flat—the sign she had been trapped, spayed, and released again.

Again I tried to go outside. Again she was gone.

I put out breakfast the next morning. She came back. Over time, she stopped running, first staying a few yards away and then waiting for me every morning. I began to call her Jane.

I knew that stray cats weren't really good for anyone—local wildlife or the strays themselves. Fed or not, they ate birds. I had seen Jane myself, hopping over the fence with a sparrow in her mouth. Cat poop littered the neighborhood (and my yard—no bare toes in the grass for me). They carried fleas and worms and infected each other with diseases from feline immunodeficiency virus to rabies. I also knew they were in constant danger from local coyotes. One of the biggest threats they faced, though, was people driving cars, oblivious to small, furry mammals.

But the damage was done. I was running a cat diner. Soon, Jane brought three more female cats. Two seemed to be adolescents, growing into their paws. A runt bumbled behind, scrawny and undersized, with a bright white mask and tall dirty-white back socks.

It became clear that Jane was the leader, if not the mother to these three. As she warmed to my presence, they did too. Soon, Jane was wrapping herself around my legs, politely asking to be petted. Her runt was soon bumping my hand with her head to be scratched as well, drooling with happiness when I snuggled her.

Jane always had the air of an impoverished gentlewoman, a high-class feline who had fallen on hard times. A cat who deserved better than this. She would sit next to me on my back porch while I read, bolt upright, with perfect posture. A lady taking a cup of tea with a neighbor.

I made plans to trap Jane and the runt with the tall white boots, take them to the vet, and then see if they might want to stay with me.

I never got the chance. In 2018 our next-door neighbors banged on our door in the July twilight. Jane had been hit by a car a block away. I grabbed a towel and raced outside.

By the time I arrived, it was too late. Lady Jane took her last, slow breaths in my arms. Her three companions crouched, unblinking, by the side of the road. I wondered what they could be thinking, watching the hairless bipeds hover over their dying comrade.

They stared motionless at us, as my husband and I wrapped Jane in the towel and carried her home. We buried her on the hill behind our house in the fading light. Afterward, we drank whiskey with the neighbors, toasting the memory of our Lady.

Jane's two biggest compatriots continued to drop by for breakfast for a few weeks. They also had the telltale trimmed ear—deemed too wild to find a home. And without Jane, they lost their tameness. After a while, they sauntered off entirely, no doubt in search of better cat food from other houses.

Months later, though, Jane's smallest runt of a kitten remained. She didn't seem to know what to do without Jane to lead her. Her tall dirty-white back socks reminded me of Julia Roberts's thigh-high boots in *Pretty Woman*.

She took up a plaintive position in our kitchen window, where she could look in to see if I was home. She begged constantly for attention with tiny, kittenish wails that pierced the glass. I started spending more and more time outdoors with her, finally training her to walk into a cat carrier for treats. One day, the treats were a trap—we went to the vet, and the cat came inside.

H. H. Boots, Purrveyor of Dry Goods (the *H*s, both of them, stand for Hooker), is now a treasured family member. She's gone from timid, tiny stray, to plump and bossy. She's sitting in my lap as I write, alternating between nuzzling her head under my arm and gazing up soulfully into my face, drooling slightly as she purrs.

Her breath is awful. I love her desperately.

A few years later, one of her sisters also came inside—and after a two-thousand-dollar surgery for severely infected teeth, added herself to our little family. The last of Jane's entourage remains at large, with other, new strays as its companions. They all get food and water from several neighbors, one of whom even maintains an outdoor cat house. I keep an eye out for them every morning on my runs. Often, I spot them lounging on the porches of the other neighbors. They cuddle with no one and are as likely to slash out at my offering hand as to run away in fear. In the evenings, I see their eyes glinting in the dusk, focusing intently on a mourning dove or sparrow below the neighborhood bird feeders—something young and naive enough to be caught. Every spring, I listen to the birds on my back porch and wonder how many of them will end up silent victims to cat claws.

As humans wreak their own havoc on local biodiversity, the cats associated with them do the same. Feral, stray, and outdoor domestic cats together slaughter one to four billion birds and six to twenty-two billion mammals every year in the contiguous United States. Cats are carnivores who happen to be live prey specialists. As long as it's small and moving, it's fair game to a cat, and species doesn't really matter. Any small bird, reptile, or rodent will do. And as cats have spread around the world, they've found plenty of small prey. At least 63 species have vanished entirely into the jaws of our favorite feline. They threaten 430 more. Cats are especially problematic on islands, where animals may never have been hunted before humans arrived with small, furry predators in tow. Ground-nesting birds and local rodents are in particular trouble. Invasive species (including cats) have been involved in 86 percent of the species extinctions on islands since 1500. There, where species haven't evolved to avoid predators, and where there's nowhere to run, cats can feast.

H. H. Boots and her adopted sister are now honored pets. But their former companions—their fellow strays—are killers. And in some places they are poisoned or shot, and people call them pests.

In places like the United States, we value cats for their antics, affection, and general adorableness. But we used to love them because they were bloodthirsty killers. Cats are obligate carnivores, pouncing on many species that eat things like grains and seeds. Birds, for example. Mice. The sort of freeloaders that early grain gatherers, storing their hard-won harvest, would be very unhappy to see. Thomas Cucchi—the zooarchaeologist we met in chapter 3—

says cats first began to be associated with humans soon after mice did, around ten thousand years ago. "The cat, which is a predator of many rodents, finds a great opportunity [around humans] to have a constant supply of mice."

We allowed them free passage on our carts and boats as we spread around the world. But cats reached the height of their pre-internet celebrity in Egypt. (The ancient Egyptian word for "cat," by the way, is—roughly transliterated—"Miw." Yes. Ancient Egyptians basically called cats "meow.") By around four thousand years ago, they were being depicted on carved ivory blades used to avert misfortune. The sun god, Ra, was thought to take the form of a cat every night to smack down the snake of darkness. Later, cats were associated with the goddess Hathor and the goddess Bastet (aka Bast). Palestine and Crete had artistic depictions of cats around thirty-five hundred years ago. There's controversy about when the cat appeared in Central Asia, but we know people there were caring for kitty injuries in the early Middle Ages. Romans and Vikings were also good for cats, taking them along as they traded, conquered, and settled.

Some modern religions extoll feline talents as well. Cats are welcome in many Muslim homes and are perceived as especially clean. A Hadīth (the accounts of people who knew the Prophet Muhammad) says that the Prophet went to prepare for prayer one day, only to find his favorite cat, Muezza, snoozing on his robe. The Prophet (or, possibly, a Sufi who had a story charming enough to end up attributed to the Prophet) cut off the sleeve of his robe rather than disturb the cat's nap. One of Muhammad's closest companions was known as Abu Hurairah, which literally means "father of cats."

But just as Christianity was not great for snakes, it hasn't been that good for cats either. By the fourteenth century, Christian writers were accusing anyone they considered "heretical" of worshipping large black cats. Male writers also compared cats to women, and never in a positive way for either party.

Cats have been worshipped, honored, hated, and hunted. But while the cat's status changed, the cat did not. Cats do what they've always done: hunt small animals, curl up in warm places, and make more cats. What changed was our religion, our culture, and our perspective.

ONCE, THERE WAS a deer mouse. Like most deer mice, it had a brown back, shiny round black eyes, and a soft white tummy. It was *Peromyscus guardia*, or the Angel Island deer mouse, and it lived on Isla Estanque, a tiny island off the coast of a bigger island, Isla Ángel de la Guarda, which is a national park in Mexico.

P. guardia also lived on Isla Ángel de la Guarda and the other tiny islands nearby. In the 1960s, scientists had no problem trapping deer mice on Isla Estanque. But when scientists came back in 1998, their traps came up empty. Empty traps also greeted them in 1999 and 2001. They couldn't find a single mouse.

What they could find was a cat, and one hundred of her bowel movements. Digging through the crap showed what the cat had been living on—93 percent of the scats contained deer mouse hair, and 2 percent had bones. The cat was trapped and, ahem, "removed" in 1999, but it was too late for the deer mice. A single cat had wiped out the entire population of the island.

Isla Estanque was a sign of things to come. The larger Isla Ángel de la Guarda also had a cat population. Now, no deer mice have been seen on both of the islands for more than thirty years. The last known example died in captivity in 1998. The deer mouse has been listed on the IUCN Red List as "critically endangered." Hope springs eternal, I guess.

It's hard to get people to care about a mouse, though. If anyone has heard of a species that a feline drove to extinction, it likely isn't the deer mice of Isla Estanque. Instead, they're probably thinking of the Stephens Island wren.

The wren, also called Lyall's wren (*Traversia lyalli*), was a small, flightless songbird that was not actually a wren (taxonomy, it's complicated). The wren was brownish, with a grayish to yellowish underside. It skittered over the ground and hopped from branch to branch like a small, feathery mouse. It was surrounded by other birds, scrub, and bushes on Stephens Island, off the coast of Aotearoa/New Zealand. That is, until a lighthouse was built in 1894, and people arrived to keep it lit.

The story goes that a lighthouse keeper's cat (its name was Tibbles) started bringing home dead birds. The lighthouse keeper, David Lyall, recognized a new species and handed over specimens to naturalists. The scientists agreed that it was a new species, but by 1895 the point was moot. In just a year, the bird was gone. An entire species, eradicated by a single, nefarious cat.

It's a sad story. It's also probably wrong.

Tibbles was not alone. There are reports of a cat being brought to the island as a gift and of a pregnant cat—that may or may not have been Tibbles—being brought by accident (it was supposed to be drowned in a sack on the way, but a storm came up and the sailors forgot). By 1897, the head lighthouse keeper complained that there were so many feral cats on the island "it would be advisable to employ some means to destroy them." The keepers pulled out their shotguns and received a bounty of one shilling per cat. The hunters followed the hunted, and cats disappeared from the island by 1925.

WE CAN GLARE at cats and tell them no. We can haul out our shotguns when things get dire. But these extinctions are really our fault. We are the ones who brought cats to all their new destinations, opening their eyes to novel, succulent cat snacks. You might think at this point that we'd have figured out how to fix this problem. As the Newell's shearwater and the Hawaiian petrel attest, we have not.

The Hawaiian name for the shearwater is the 'a'o, a reference to its call, which sounds like what might happen if a very small donkey learned how to giggle. Despite its scientific name (*Puffinus newelli*), the 'a'o are not puffin cousins (taxonomy, why). They're kind of your basic seabird, with a dark gray back and white stomach. Their heads are also dark, giving them the visage of a bird that sees its looming extinction with all the gravity the situation requires. Scientists thought the 'a'o had been hunted to extinction by 1908. Then they discovered breeding populations of the 'a'o on Kauai and hopes for it recovered.

The Hawaiian petrel (*Pterodroma sandwichensis*, a reference to the Sandwich islands, which is what James Cook called Hawaii) looks a lot like the 'a'o. It's got some brown on the top and a whiter face, and its expression is a little more lighthearted. Its Hawaiian name is the 'ua'u, a reference to its haunting, hooting call, the sound that might be produced by the result of a mourning dove and an eagle owl's ill-conceived romance. The 'ua'u is known to nest in Haleakalā National Park and on Mauna Kea, Mauna Loa, and Kilauea on Hawaii.

The 'ua'u nests in cliffside burrows, while the 'a'o prefers to burrow under the ferns that cover the slopes and cliff edges. They both nest from early spring to late fall and spend their winters out over the sea.

Hawaii cares deeply for its wildlife. So deeply that the whole state converted to digital TV in January 2009—one month earlier than the rest of the nation. The reason? The analog towers needed to be taken down before the 'ua'u arrived for breeding season.

But digital TV didn't save the birds. Right now, the IUCN Red List estimates that there are fewer than twenty thousand 'a'o left, and fewer than eleven thousand 'ua'u. The numbers seem large. But they're not enough. Both the 'ua'u and the 'a'o probably mate for life, and both have only one egg per year. The populations aren't growing anytime soon. Right now, they're actively shrinking.

Why? Cats, at least in part. Some of the breeding colonies left to the ʻaʻo and the ʻuaʻu are in the Hono O Nā Pali Natural Area Reserve. It's not an easy area to get to, says Alex Dutcher, a co-founder of Hállux Ecosystem Restoration, which works to manage invasive species on Kauai. "The dominant vegetation is this Uluhe fern," she says. "It grows in these huge, tall, thick, dense mats. You don't know where the ground starts or where the cliff starts. It doesn't cut with a machete. It's really horrible. You have to crush your way through it. It takes forever."

Dutcher's group spends a lot of time smashing through the brush to manage cats. Where humans stomp and crash, cats slink without a care. In a survey conducted from 2011 to 2017, André Raine and his colleagues at the Kauaʻi Endangered Seabird Recovery Project found evidence of 309 depredations. Half of those were due to rats, 10 percent to pigs, and 4 percent to barn owls. The remaining 36 percent were due to cats.

It might seem like rats were the biggest problem here, but no. Rats attack and eat the eggs and young. Cats attack and eat the adults. Take out a chick or egg, you take out a year. Take out a breeding adult, and you lose years and years of potential eggs. Raine and his colleagues estimated that if predators were not controlled, the ʻaʻo and ʻuaʻu would be headed for extinction within fifty years. That seems like a long time, but it's all too short. Think of the dodo, the passenger pigeon. The Stephens Island wren. We knew they were becoming rarer and rarer, yet we did nothing. If we could go back in time and prevent the extinction of the dodo, wouldn't we?

Determined to protect the birds, Dutcher, Raine, and their colleagues are part of a human fence around the remaining ʻaʻo and ʻuaʻu. The goal is to track and remove as many cats as they can from the Hono O Nā Pali Natural Area Reserve. They've seeded the reserve with game cameras, which are checked multiple times a week. Speed is critical. When a cat (or rat) has been near a burrow, "we know right away," Dutcher says. Then, they head out into the reserve to place live traps. "We use lots of baits and lures to entice cats in," she explains. "We use a mix of feathers or shiny objects, we use scent lures with herbal oils mixed with musk-type scents and fishy-type baits or meaty baits. We keep things pretty mixed up. We find that by rotating through different things we get all cat preferences."

In the reserve, cats are not the cuddly, sweet, funny creatures that sit in our cardboard boxes and ride our Roombas. The cats they are after, Dutcher notes, are not even your average stray, used to people and pet food. "Most of the cats we capture in the reserve have never seen people," she says. "They don't meow at you, they don't look at you, they don't see you, they just see you as something to get away from."

The cats they catch won't survive their first human encounter. In order to save the birds, the cats must die. If conservation of threatened, wild, defenseless species of birds that exist nowhere else in the world is important, these cats are threats. They are pests.

It's a hard thing to contemplate for a lot of people. It bothers me, sitting with my pampered, purring cat in my lap. But then, I mentally replace the word "cat" with "rat." Suddenly, I'm less upset—and I'm a person who likes rats.

Despite her job, Dutcher loves cats. She has two indoor cats of her own. "If I didn't have to kill another cat for the rest of my life, because there were none outside, I would be thrilled," she says. "Nobody likes killing cats; it's protecting birds that we like doing." Dutcher blames society for allowing cats to get out of control, for seeing cats as a species that can be left to its own devices.

DUTCHER IS NOT the only person who combines a love of cats with working to get rid of them. Sarah Legge is a conservation biologist at the Australian National University and the University of

Queensland. She's also on the Australian Feral Cat Taskforce. "I used to rescue stray cats when I was a kid," she says. "And my first job was on Scottish wildcats. I've worked on lions—I've always had a thing for cats."

In a 2020 parliamentary report, Legge and her colleagues estimated that Australia is home to 2.8 million feral cats—that is, cats without any clear owner. While that sounds like a lot, it's nothing compared to the United States, which plays host to a (very roughly estimated) 30 million homeless felines.

But numbers, Legge says, don't say anything about impact. And in Australia, the impact of feral cats is massive. The First Fleet of white settlers in 1788 probably brought cats to what would become Sydney. For twenty to thirty years, cats (and colonists) established themselves in and around the new city.

Then, a catsplosion took place. Over the next seventy years, cats established themselves across the whole of Australia. "As settlers expanded, they'd take cats with them," Legge says. "But cats got ahead of that wave." And the cats ate. And ate. And ate. Legge's parliamentary report notes that cats are a major cause in the extinction of twenty-five Australian mammal species from the time they arrived in 1788 to 2020. In response, Australia has declared war on the feral feline.

To protect their native species, Legge says, Australia is prepared to try a lot. The most effective way, she explains, might be fenced areas or offshore islands, places where cats can be firmly kept out. These havens have prevented the extinction of thirteen mammal taxa, Legge says, "sometimes by accident." In a few cases, species that were hunted to extinction on the mainland still have isolated populations on islands—islands that so far have no cats.

But fences and islands are small. And when you confine an animal to those specific areas, Legge explains, you lose the function of that animal in the rest of the ecosystem. It's nice to have it there, of course, but it will no longer be playing a role in the greater environment, no longer digging burrows or dropping guano or adding the many things that animals add to ecosystems they live in.

Poisoning, shooting, and trapping are also used, and Legge notes that some Indigenous Australian groups have developed expertise in cat hunting, including the Kiwirrkurra traditional owners in the Gibson Desert. "Killing anything is unpleasant, but you have to weigh up the consequences of acting and not acting," Legge says. "If you don't do that, not only will a lot of other animals suffer, but they'll suffer to the point where extinctions are plausible. Of course, it's a very distasteful thing to kill any animal. But not doing it is worse."

There are other ways than killing. Sarah Legge's research has shown that promoting plant ground cover—shrubs, bushes, and grasses—helps native prey escape from ravenous cat jaws. But when I asked about TNR, Legge's cheerful manner exploded into frustration. "There are trials under way right now," she says. "It gives me fits."

TNR, or trap-neuter-return, is the method promoted by many animal welfare groups in the United States. It's exactly what it sounds like. Feral cats are trapped, they are spayed or neutered, and then they are released back where they came from. Sometimes, one ear is trimmed flat at the tip (as I saw with Jane), to indicate that the animal doesn't need to be trapped again. It seems like a good, humane strategy. The cats live out their lives, and the colony eventually dies out. Right?

Julie Levy, a veterinary scientist at the University of Florida, certainly thinks so. She's been practicing TNR herself since she was in veterinary school at the University of California, Davis in the 1980s. This was before the internet was really a thing, she says, so "people were discovering TNR on their own. . . . Everyone was inventing it in their own way. Our first attempt was to go around trying to catch these cats by hand, chasing them around with a net and getting bitten."

Since then, Levy's life's work has been starting and running large-scale, low-cost TNR clinics, one in Raleigh, North Carolina, and the other in Gainesville, Florida. Both are called Operation Catnip and combine large TNR programs with surgical training for veterinary students. Operation Catnip of Gainesville boasts more than sixty-five thousand sterilizations since 1998.

In the process, Levy has kept track of how well TNR programs do at reducing cat populations. In a 2005 paper looking at TNR programs in San Diego County, California, and Alachua County, Florida, Levy and her colleagues built a population model based on their TNR data. Over a decade, cat populations never stopped growing; just as many female cats were pregnant as before. But the model showed that population reduction was possible, if you could TNR between 71 and 94 percent of the cats. A 2022 study out of Israel made the model a reality. The scientists showed that after neutering about 72 percent of the cat population of Rishon-LeZion, a city south of Tel Aviv, cat numbers started going down by 7 percent per year on average. But even that decrease had limits. New cats were constantly arriving, and the cats left to breed bred even more.

Reducing the population requires constant vigilance and constant cat snipping. Many stray-cat colonies are dumping grounds for people's unwanted (and often fertile) cats. The colonies are fed by kindhearted people or by accident, and food brings more cats. "If there's a food source, the number of cats is not unlimited, but it can get very crowded," Levy says. TNR groups are constantly neutering, and new cats are constantly arriving.

Pete Marra, a conservation ecologist at Georgetown University, and author of the book *Cat Wars: The Devastating Consequences of a Cuddly Killer* (a bit of a giveaway of his point of view), is quick to point out how impossible the math really is in places like the United States. Suppose, he says, that there are thirty million cats in the United States, and six million of those are already neutered. That leaves twenty-four million, half of which would be females that could make kittens. Even the Humane Society of the United States says that's optimistic, and that only 1 to 2 percent of the feral and stray cat community has been sterilized, or about six hundred thousand. Even with Marra's more optimistic numbers, "say [each one] produces four more cats. They're shoveling sand against the tide. . . . A giant, furry, meowing tidal wave killing everything."

Levy is not nearly as pessimistic. "There's been a flurry of papers that show it's realistic," she says. One of them, by Daniel Spehar and Peter Wolf, updates the results of a long-term TNR program at the University of Central Florida in Orlando. In 1991, cats were everywhere, and the university was putting in a lot of calls to pest control. "And so the staff actually just kind of . . . assigned themselves to start doing TNR," Levy says.

Being a campus full of academics, the faculty and staff didn't just go around trapping random cats. They mapped colony locations, maintained feeding stations, carefully kept track of the cats and their health, and took meticulous notes. The campus itself contained about sixteen individual cat colonies, which over a twenty-eight-year period contained more than two hundred cats. Over time, the staff managed to TNR 85 percent. By 2019, there were only ten cats left.

But in all my reading of TNR papers, two major things jumped out. In most successful studies, about half of the cats ended up adopted. "Even if the goal is TNR, there's nearly always this organic adoption process," Levy says. The other is that the programs take decades to produce results. In Levy's 2005 modeling study, the cat populations continued to grow. In the successful UCF study, it took twenty-eight years for the population to shrink down to its last ten cats.

If the TNR'd cats are returned to a colony, they live their life out. If they are in a fed colony, that life will be a lot longer than if they are facing nature red in tooth and claw. Some of the TNR'd cats at the University of Central Florida lived more than ten years, much longer than the average outdoor life span of between two and eight years. Every extra year is potentially a year of hunting, a year that some endangered species can ill afford to lose.

Back in Australia, TNR is currently against the law in most states. Cats are domesticated animals, Legge says, so letting them go is abandonment, meaning that the "R" in TNR is illegal. But some groups have been pushing for TNR as a solution to Australia's cat issue, and some small programs are under way.

Legge is afraid that introducing TNR into Australia will not only be futile, it will make people resist the other, less publicly palatable options. "It's like climate denialists," she says. "You shouldn't give them the airtime."

That seems harsh. Many TNR groups do work tirelessly to keep as many cats as they can, taming them and trying to find them homes or jobs as barn cats. They acknowledge that outdoor cat numbers should be minimized as much as possible. And many people in the United States, at least, have been told that TNR works. They may not have seen the studies themselves, but they trust adoption agencies and nonprofits.

Large pet charities are pretty convinced that TNR is the right thing to do. PetSmart Charities and Petco Love have both offered funding to organizations that perform TNR, and they pay for future veterinarians to be trained in it. In the process, they help spread the gospel that TNR is the most effective and humane way to deal with stray-cat populations.

Poison, shooting, even TNR. If this seems cruel—well, it is. But imagine an analogous invasive species arriving on the shores of North America. Imagine a new predator slinking ashore to find defenseless prey, and dining on mammals and birds alike. Imagine forests and swamps gone silent, ecosystems destroyed.

In fact, we don't need to imagine. The story of the Burmese

pythons in the Everglades offers a similar tale with a reptilian twist. There, we have no hesitation conducting python hunts. We aren't bothered by the idea of trapping or killing the snakes. The problem is clear, and so is the solution. The snakes are invasive. They need to go.

On the other hand, why does the fact that something is wild and native to a place make it more important in our eyes than things that are not? Species move all the time, but to humans, some movements are allowed and some are forbidden. When they are allowed—as with the shifting of a salamander species to the north because climate change is making their former habitat untenable and a new habitat possible—we call them "climate tracking" or "climate refugees." We say the clever, adaptable animals are taking advantage of a new habitat. That an ecosystem will adjust. When animal movements—and their effects—are not what we want, however, we call them "invasive species" or "pests."

In some areas, TNR might be enough—in suburbs and woodlands, places where the birds fly and the small mammals learn quickly about the perils of cats. Whether we want it or not, however, cats on islands like the Hono O Nā Pali Natural Area Reserve can cause seabird destruction. It's destruction that ground-nesting birds can't escape. So maybe, as Emma Marris notes in *Wild Souls*, we need to be careful what we tar with the brush "invasive species." "Instead of a paradigm where we see all 'foreign' species as malevolent invaders that should be considered threats to ecological integrity unless proven otherwise," she writes, "maybe we should instead see island species as particularly vulnerable to newly arriving species." The cats are strong. But the birds are also weak.

AUSTRALIA. MEXICO. STEPHENS ISLAND. HAWAII. These are extremes, examples of what can happen when cats roam free in a defenseless landscape. But these islands aren't suburban backyards in North America, abounding with squirrels and sparrows. They aren't farms with barns and mice. That cats kill is certain. But what they kill, and whether that matters for the ecosystem as a whole, is a bigger and tougher question.

Marra and his colleagues estimated in 2013 that cats kill 1.3 to 4 billion birds and 6.3 to 22.3 billion mammals per year in the continental United States. But Wayne Linklater, a wildlife biologist at California State University, Sacramento, doesn't have a lot of patience for estimates. "I'm bemused when our colleagues try to produce numbers like that," he says. "The real purpose of those numbers is flag-waving. They are political numbers more than they are scientific numbers. We produce a number because it communicates something startling to the public, not necessarily because we think those numbers are particularly accurate."

Which cats hunt and what damage they do might vary a lot from ecosystem to ecosystem, Linklater says. Marra's paper estimated, for example, that feral cats carried out most of the carnage. But feral cats aren't the only happy hunters. Ask anyone who owns an indoor/outdoor cat what Fluffy has brought home, and you'll hear tales of perfectly alive squirrels, half-dead lizards, and dead bats left to rot in places just out of human reach.

Roland Kays, a wildlife ecologist at North Carolina State University in Raleigh, conducted a citizen-science-assisted study of owned cats, published in 2020. He handed out 925 GPS collars to cat owners in four citizen-science projects in the United States, the U.K., Aotearoa/New Zealand, and Australia. The owners attached the collars to their cats and filled out a survey about how much prey their furry friends brought home. From the GPS data and the surveys, Kays and his group calculated how big a territory was for the average house cat. They also calculated how much prey those cats were killing each year.

It turns out the average territory of most cats was about 3.6 hectares, or about five to six soccer fields in size. Proportional to a cat's body size, that's huge, but to a human, it's just a few single-family suburban homes. With those numbers, and the average calories a cat needs per day, Kays says, they could figure out hunting success. "We standardized it based on how many calories are in a mouse," he says. (It turns out that someone really did put a mouse in a bomb calorimeter to find out how much energy it contains. They're about forty-one calories each.) Based on the calculations and the survey data, Kays concluded that pet cats allowed outside were killing between fourteen and thirty-nine prey per hectare, per year.

Because house cats are fed, they're probably not hunting as much as feral cats. But fed cats still kill for sport. It could be because the processed kibble many fed cats receive isn't high enough in protein and they're trying to get that delicious, delicious meat. It could also be that while humans and cats have been associated with each other for a long time, humans haven't generally been feeding most of the cats they're associated with Fancy Feast throughout history. Cats had to find their own food, thus keeping their hunting instinct intact. Finally, cat hunting instinct and cat eating instinct aren't the same thing. It could just be evolution—you end up with more available food if you keep the hunting hustle going.

"Cats are likely having very high ecological impacts on prey species, sometimes five or ten times what [one] would expect from a native predator," Kays says. "They live at higher densities and kill more." Cats aren't just leaving you something unpleasant to step on in the morning. They're impacting the biodiversity where you live. But because of their territory size, Kays explains, "the good news is it's concentrated near the house." So at least it's only very local biodiversity.

The numbers look bad. But numbers aren't everything. "Conservation scientists want to show how bad cats are," explains Sarah Crowley, an anthrozoologist (someone who studies people and animal interactions) at the University of Exeter. To do that, scientists published big numbers to show how much cats kill. But without context, she says, those numbers just aren't very useful.

They don't, for example, tell anyone the nature of the prey that cats are killing in many areas, Crowley says. Are they killing the old, weak animals? The ones who wouldn't make it anyway? Or are they dealing death to things that might otherwise survive? Are they killing other invasive species? Or native species that play important ecological roles and can't defend themselves?

If an urban or suburban cat in the United States is spending time hunting, for example, their prey will be common wildlife— voles, moles, squirrels—not endangered species. Some might be other species that humans have imported—pigeons, mice, sparrows. (Interestingly, rats appear not to be a hot menu item. In a 2018 study of a cat colony near a rat-infested area, Michael Parsons, an urban ecologist at Fordham University in New York City, found that, mostly, rats and cats just avoided each other. The rats rarely ended up as dinner. It's probably self-preservation, as a rat is large enough to do significant damage to a cat—damage that doesn't make the takedown worthwhile. But honestly, kitties, you had one job.) These prey reproduce very quickly, not like the slow-living species on islands. Cats aren't going to drive any of those populations to extinction anytime soon.

Crowley agrees with Linklater that big whopping kill numbers aren't "particularly valuable in moving the discussion forward," she notes. "It doesn't tell you how to fix it." What might do it, she says, is a better idea of which cats kill and which don't.

Crowley and her colleagues surveyed 219 feline households, asking how much their cats hunted, and instructing owners to give the cats opportunities to reduce hunting with play and food. In the survey, Crowley says, there were two cats named Jeeves. One of them took after his namesake—the ridiculously efficient butler from P. G. Wodehouse's books. "One was a crazy hunter; in a three-month trial he killed more than one hundred fifty things," she recalls. If you took Jeeves's kills and extrapolated them to all the cats roaming the planet, every rodent and bird would be facing its doom. But Jeeves was an outlier. Most cats never brought home prey in those numbers, and some never brought home anything at all.

BEFORE WHITE COLONISTS arrived in Aotearoa/New Zealand, the only native mammals in the country were bats and marine mammals such as seals. There wasn't a rat, cat, or possum to be seen.

Separated from mainland predators for about eighty million years, many of the local birds and reptiles took up lives of leisure. A significant portion of birds just gave up on that whole flight thing. Species like sparrows that lose a lot of members to the gullets of predators tend to reproduce early and often. But if no one's eating you, you can afford to take your time. Many birds in Aotearoa/New Zealand live long lives. They wait to breed, laying few eggs when they do.

Rats hit the islands with the first Māori settlers. Then cats sauntered ashore in 1769 with Captain James Cook, soon joined by stoats and possums. All were probably very pleased to find snacks sitting tamely on the ground, unable to do more than waddle awkwardly away. Species quickly began to decline and then to go extinct.

The Aotearoa/New Zealand government now maintains a web page devoted to the damage of introduced predators. On the feral cat page, a dead cat lies next to the remains of 107 short-tailed bats. It is strongly implied that this cat took down those bats. The cat is a tabby, with a white belly and tall white boots on its back feet. It bears an unsettling resemblance to the cat in my lap.

At the same time, New Zealanders adore their cats. About 41 percent of the population own pet cats (compared to between 24 and 38 percent of American households, depending on which interest group you ask). "The problem is a population problem . . .

a lot of cats in a small area," Linklater says. Some still want to use cats for their original job—pest control. In 2021 the town of Queenstown started a "working cats" program, letting out TNR'd cats onto rural properties to manage rabbit and rat populations, apparently without any hint of irony.

Aotearoans/New Zealanders really don't believe in keeping cats cooped up. Cat owners related to their cats' independence, Linklater explains. "There's a strong belief that the cat needs to roam and be wild. It's quite different from our relationship with our dogs." People also didn't want to believe that their cats were part of the problem, he says. "It's clear people underestimate how much depredating cats do."

Even veterinarians in New Zealand don't want to see cats cooped up. In a study that Linklater and his colleagues published in 2019, they found that 67 percent of Aotearoa/New Zealand's veterinarians believed that keeping cats inside twenty-four hours a day would have a negative effect on the cat.

I glance down at H. H. Boots. Fast asleep now, she's warm, and her nose is adorably pink. When awake, she seems like a hopeless hunter. If a moth or a fly gets in, she'll watch it for a while, then leave someone else to deal with it. Her preferred hunting prey are licorice-flavored Brach's Jelly Beans. They roll satisfyingly across the floor—and never fight back. I can't imagine her successfully killing and eating anything.

And yet, though I don't like to think about it, she probably did. Her maybe-mother Lady Jane routinely leaped through my yard with the remains of a sparrow. Her sister still waits under bird feeders for doves to get a little too comfortable. My sweet little kitten probably was a killer.

Keeping cats as pets and denigrating them as pests also doesn't acknowledge the work that cats do. In her experience growing up in a Muslim country, says Sarra Tlili, who studies and teaches languages and literature at the University of Florida, cats weren't "pets in the traditional sense." Or rather, not in the sense that many people in the United States probably think of pets—with toys and catios and heated beds. They are welcome inside, but they aren't expected to really live there. Windows and doors are kept open for airflow, and as air flows, cats flow too. They also have a job—to catch mice and cockroaches and rats.

Cats merit several mentions in the Hadith. "There was a woman who had a cat," Tlili says. "She kept it at home and did not feed her and the cat died. The Prophet said the woman is punished by hellfire because she did not allow the cat to go out and feed [itself] and she did not feed the cat herself." The mistreatment of the cat stands in for mistreating animals more generally.

"They are perceived as a family member," Tlili explains. "My own aunt, if the cat is hungry, she will give her own piece of meat to the cat. If the cat is wounded they feed the cat, if it needs help they will provide the help." But it's not a pet.

Cats don't just deserve care, they deserve respect as working members of the family and also as individuals. Tlili says that her interpretation of Islam means that cats should retain their reproductive freedom. TNR is "an act of violence against animal bodies that we don't have the right to do," she says. But Tlili isn't everyone. There are as many interpretations of Islam as there are of any other religion, and other observant Muslims feel just fine fixing their cats.

When cats kill local wildlife, though, traditional religious texts—Muslim, Christian, or otherwise—don't have much to offer. "The issue of biodiversity is never addressed in traditional texts and those are the main texts I read," Tlili says. "I feel that the answer from those texts would be [that] taking care of biodiversity is not . . . humans' business." Islamic tradition has a lot to say about helping individual animals in distress, and not cutting down trees unless people need them. But most ancient traditions didn't focus much on humanity's impact on biodiversity.

Maybe the job is beyond us. "I do trust that nature is capable of regulating itself, that God is capable of bringing back that balance if you're thinking in theological terms," she says. "Humans have always been the most problematic species on earth, this is not a new specialty we acquired." Tlili also notes that when we try to fix the mistakes we've made—killing wolves, introducing cane toads, slaughtering sparrows—the results often come back to bite us. We never know quite enough not to screw it up.

PEOPLE LIKE TO create dichotomies. Republican or Democrat. Cat person or dog person. Views of cats, though, much like cats themselves, gleefully buck this trend. Pet? Pest? Working animal? Cats can be all of these to us—often at the same time, Crowley has found. She had fifty-six U.K. cat owners perform something called a Q-sorting. They took a whole stack of index cards with different statements about cats and ranked them, based on how much they agreed with them. The result is a different bell curve of belief for each person.

Based on that, Crowley was able to divvy up U.K. cat-owning beliefs into five groups. The first are the "concerned protectors," the helicopter parents of cats. To these owners, cats are in constant danger outside. They could get hit by a car. They could get eaten by . . . something (though in the U.K. is there anything left to eat them?). They could try drugs or take candy from a stranger. Concerned protectors felt good keeping their cats in, at night at least, and sometimes all the time, especially if they could provide entertainment. They didn't worry about the birds at risk from their little hairy darlings. Their concerns were all about the cat.

The second group were the "freedom defenders," the free-range cat parents. They saw cats as wild animals with needs, needs that could not be fulfilled with kitty TV or catnip-filled mice. For these owners, hunting was a benefit. It was a sign of a healthy cat and free pest control.

The third through fifth groups were mixtures of the helicopter parent and the free-range. Some saw cats as wilder pets, where time outside had more pros than cons. Hunting wasn't great, but it was just cats being cats, you know? Others still liked cats to roam but were even more worried about hunting. Not worried enough to do anything, but it's the thought that doesn't count.

CATS WILL ALWAYS be cats. Reducing cat predation on native species, then, will mean changing the hearts and minds of people—the people who let their cats hunt. "The thing that's really important is to start looking for guidance for cat owners that's tailored not just to their viewpoint . . . but to their circumstances," Crowley says. Where is the person located, are they able and willing to spend time giving their cat things to do, things to watch and play with?

And what about the cat? Some are Jeeves, reckless slaughterers. But others are not. Based on the data, Crowley notes, "a lot of the cats don't really hunt at all. They bring in one thing in six months. For those people, they're not going to worry too much." We should be focusing on the really good hunters, she says, the cats like Jeeves.

In his study, Linklater asked cat owners and veterinarians what they might be willing to do to try to reduce cat predation in New Zealand. Many were willing to put a collar on their cat. Few were willing to confine their cat for twenty-four hours a day. "I think as conservation biologists we sometimes forget that the answers we want people to adopt are either unrealistic or take a long time," he says. "Changing people's values and beliefs is not something that happens quickly."

Especially not if cat owners feel alienated by studies saying cats kill billions of birds. "All it does is aggravate cat owners," Linklater says. "It doesn't seek to have a conversation with them or understand them." He hopes that by engaging in a conversation, and maybe keeping cats inside a little at night, cat owners might be willing to take further steps to keep their little predators indoors.

Cat owners can even decrease hunting behavior by giving the cat what it wants: meat and activity. Crowley and her colleagues found that five to ten minutes of "object play" (wands with feathers, for example) and a high-protein, grain-free diet decreased the prey brought home by 25 and 36 percent, respectively—even though the cats were still being let out.

Kays agrees that there are ways to let the cats out that might make them less deadly. "Our take-home is keep your cat indoors," he says. "But a corollary is if you're not going to do that, try some methods to have them hunt less." Bells on cat collars might help (though some cats can figure out how to move without jingling), and brightly colored collars may make them more visible. And bonus, Kays says, it makes the cats look very grumpy.

ONE THING CAT lovers and cat haters can agree on: it's better to have fewer stray cats. So why do some people think TNR is as evil as climate change denial, while others send hate mail and death threats if cats get euthanized?

"You can agree on the outcome but not agree on why you're doing it, and then when you get into how to do it you come to loggerheads," says Kirsten Leong, a social scientist at the National Oceanic and Atmospheric Administration, based in Honolulu.

Leong has been studying how people see the animals around them. In a 2020 paper, she and Ashley Gramza, a conservation social scientist with Arkansas Game and Fish Commission in Little Rock, explored the cultural models that different groups use to think about cats. "Cultural model" just means an assumed idea within a particular social group. In Chicago, for example, it's assumed that deep-dish pizza is pizza. In New York City, such an idea is heresy. These are cultural models of what "pizza" is.

In this case, the social groups are wildlife conservationists and animal welfare professionals. They live in the same state or county, sometimes in the same town. They might even be members of the same family. But they have different cultural models when it comes to cats. Modern wildlife management in the United States, Leong explains, arose out of game management—the management meant to promote animals we like to hunt. A lot of wildlife management culture did too. It became about protecting species that humans find valuable—by whatever means necessary. She says that you have to think about "the perspective and set of beliefs you need to be seen as credible in the profession." One of those beliefs is that cats are an invasive species. And when managing invasive species, "lethal" population management is one of the tools they use. Python or pussycat, it shouldn't matter.

But for animal welfare groups, "invasive" is nowhere in their concept. To them, stray and feral cats aren't pests, they are homeless pets. That cultural model is pretty widespread in the United States, Leong notes: "If you Googled 'cats' you wouldn't see them predominantly represented as invasive species." The dominant idea of cats as pets, she says, is because that's how most people encounter them every day. Very few people in the United States see cats stalking endangered species through the brush.

That dominant view could change, though, as people's experiences with cats change. Leong points to the example of Australia, where cats are designated pests. You can hunt and kill a feral cat anytime you want in Australia.

That attitude might elicit a gasp of horror from someone snuggling a cat in the United States. But it also shows just how differently Australians experience cats. Australia is an island and is intensely proud of its wildlife. Kids are taught all about their important native species that are found nowhere else in the world. They are also taught about invasive species and the dangers they pose.

The cats themselves? They're the same animal. "They can

adapt," Leong says. "They're just doing their thing. But the longer [invasive species] becomes our main experience with them, the more they become stigmatized."

We like to think of scientists as immune to bias and allergic to emotional argument. But they are on either side of this cultural model, too. Marra—who estimated bird loss in the billions—sees TNR as hopeless and notes that while people study the effects on the cats, they don't study the effects on the local wildlife. "The last thing [TNR advocates] want is for people to go out and study this, they don't want to know the answers. People have cats in barns all over the place, and they say if they didn't the rodent population would go crazy." Marra wants to challenge that, but says no one will give the guy who wrote *Cat Wars* access to a TNR colony. I can't say that I'm surprised.

Marra says he received death threats when his kitty-bashing book came out in 2016. (The irony of people who hate the idea of killing cats threatening a scientist with death escapes neither one of us.) "They don't want to believe that cats are a problem," he says. "They want to believe their gut."

Levy, for her part, wants to know why the burden of proof continually rests on the side of TNR, and points to the studies that show, in the long term, reductions in colony population. "People who ask us to do that don't ask whether culling works," she says. Levy wants to see the list of papers studying the impacts of culling compared to the list of papers studying the impacts of TNR, to see which really does produce the bigger effect. Culling, she explains, only takes care of cats that are there, not the ones that could move in. Take the cats out, sure, but as long as there's a food source (and it's not an island), more will arrive. And, she says, people will volunteer to help TNR, "but no one has yet been able to recruit a volunteer force" to kill cats in the United States.

"What I think is a false narrative is that there's a dichotomy, that you have to choose one or the other," she says.

INVASIVE PEST? OR homeless pet? Both of these can be true, at the same time and even in the same place. If we believe cats are pests, if we label them killers, cats will die. If we label them a homeless pet, to be fed and let free, species in some places will continue to go extinct. Until we can hold both of those concepts in our minds, we will fail both to effectively prevent the harm cats cause, and to effectively take care of an animal that many of us care deeply about.

Even after reading about billions of birds dead and species gone the way of the dodo, it's hard to look at H. H. Boots and see a killer. So many of our ideas about pests are about what we see and believe, our ideas of what an animal is and where it's supposed to belong. When a mouse invades the house, we feel a loss of privacy that is sudden, intimate, and shocking. If our house cat kills the mouse, all the better.

When a cat drives a wren to extinction, we are sad. But it's the distant sadness of something that is happening half a world away, something for which we bear no responsibility. But if cats are a threat to the biodiversity we say we value, then yes, cats are pests. And the responsibility for that lies squarely on us.

When I looked at Lady Jane that first time, slinking slow and skinny through my yard, I remembered a line from Antoine de Saint-Exupéry's *The Little Prince*, where the title character meets a fox, and learns what it is to tame animals and to develop ties with other creatures. The prince eventually leaves the fox to continue on his travels, but the fox leaves him with this: "'Men have forgotten this truth,' said the fox. 'But you must not forget it. You become responsible, forever, for what you have tamed.'"

Humans spread over the world, bringing their animals along, with little idea that as we claimed ownership, as we "tamed" spaces, we would end up being responsible for what happened to them. Often, even as the planet warms around us, we still try to forget what we are responsible for.

We are responsible for the cats in our lives, even the feral strays.

We are responsible both for their lives and for the deaths they cause. Whether we see them as a pet, necessary rodent control, or a deadly threat to biodiversity, they are where they are because we put them there. And now, as they threaten more species with extinction, we are faced with a choice. Who lives and who dies? Who gets honored and who gets vilified?

PART IV

THE POWER
OF PEST

7

A BAND OF
COYOTES

The door to the converted shed opens, and Niamh Quinn, last seen trying to put a collar on a rat, strides in, hauling a dripping trash bag. "It's a wet one," she says, trailing blood as she plunks the bag on the scale. After weighing, Quinn heaves it up onto the necropsy table and pulls it open, revealing the brown paws, rough coat, and battered face of a long-dead coyote (*Canis latrans*). The coyote, now known as 632, became roadkill on February 23, 2021, only a few blocks from the research station where it is now going to be cut open for science. It got picked up, bagged, and placed in a freezer, where it remained for nearly six months, until it was taken out to thaw in a wheelbarrow in the sun for about four days before my visit.

At first, the smell is faint, but as more of the coyote is revealed, the stench of death—sickly sweet, gag-inducing rot—quickly threatens to become overpowering. A student excuses herself. I keep having to remind myself to breathe through my mouth as Quinn cheerfully inspects the body.

Before she met the front end of a car, 632 had been a healthy adult female. Her fur is thick and luxuriant, her muscles strong. If I didn't know better, I might think I was looking at someone's well-loved dog, with alert ears and a fluffy tail. But she's just slightly off from dogdom. The fur is slightly too thick, the face slightly too narrow. Something about her whispers "undomesticated."

Quinn seizes a pair of clippers and begins to shave, taking fistfuls of fur and shoving it into various prelabeled bags. Another team of scientists is studying the hair, trying to see if they can detect rat poison the coyote may have ingested. Once the bags are loaded, Quinn picks up a scalpel and deftly slices off one ear for genetic analysis. Then, she grabs an old chunk of wood and places it beneath the coyote's neck. I'm reminded suddenly of the chopping blocks used when tyrannical kings send their wives to meet the ax.

I was right. It is a chopping block. Quinn slices through the muscle to the spine and grabs a machete and a mallet. A few well-placed thwacks and the head is off. It fits neatly into another plastic bag.

Quinn moves down the body, and warns me and two students that she's about to go in. We all take a respectful step back. As bodies decay, the digestion of microbes can build up gases inside an animal's torso, leading to a potential explosion if the torso is pierced in the wrong place. Quinn moves carefully, and the result is a speedy deflation and no additional spatter. The smell gets even worse, with added notes of bile, and I remind myself forcefully that I have never thrown up in a necropsy. Yet.

Quinn removes the liver, which will also be checked for evidence that 632 dined on potentially poisoned rats. With two cuts, she pulls out the entirety of the gastrointestinal tract. Feeling around, she says, "I think I feel a spine." A slice into the stomach reveals a pile of partially digested fur, and a few pale, shining bones. Maybe a rabbit. A well-digested kitten? It's too far gone to tell. Quinn bags the whole tract for a researcher studying the gut microbiome.

Rummaging around in the body cavity, Quinn pulls out something like a string of pale, glistening sausages. The four tiny bulges are in 632's uterus. She was pregnant.

Many people think of coyotes as pests. They take the form of sinister, wolflike beings that haunt our neighborhoods, attack our kids, eat our pets, or prey on our chickens. Carnivores that prove our power over the landscape is a lie. In that moment, staring at the rotting body on the table, I feel heartbreakingly sad. They, not us, seem like the powerless ones. Coyotes have succeeded with humans, and in part because of them. But what we give, our cars, traps, and guns take away.

Wolves need natural spaces and big game. Coyotes aren't so choosy. They're happy to dine on the fruit and avocados that feed the granola crunchers of Southern California. They also pounce on the livestock, rats, and stray cats that follow human settlement. They learn how to walk on dividing walls between houses, how to den in compost heaps, how to survive and thrive in downtown Chicago and even Manhattan's Central Park.

In the process, they've gone from the smaller, safer wolf that nobody feared to the pet-slaughtering terror of the suburbs. We love wild animals. But we only want them in the wilderness. Our human-dominated landscapes are another matter. There, we are in control, and the only animals allowed are the ones that are in our power. When nature reminds us that our control is never absolute, we feel it like a violation of our sacred space. We respond with anger, hatred, and violence.

These reminders of our limits—in the yipping howl of a coyote, the scrabbling of a rat, or the bear at the bird feeder—might teach us something, if we cared to listen. Coyote 632 paid the ultimate price, simply for being where she was. But she was, in fact, where she belonged.

IT'S EASY TO think of pests as having followed us, settling in our cities, our habitats, in flocks or skittering hordes. They come where we are, not the other way round.

But coyotes have always been in Southern California. Some of the more unfortunate ones ended up in the tar pits of Rancho La Brea in Los Angeles tens of thousands of years ago. As Dan Flores describes in *Coyote America: A Natural and Supernatural History*, coyote bones showed up in Chaco City, New Mexico, a thousand years ago, and coyotes played a significant role in Aztec religion.

When the Navajo see a coyote, they also see Coyote—an ancestor to coyotes and an important figure in his own right. In *Navajo and the Animal People*, Steve Pavlik, a (non-Indigenous) professor of Native American studies at Northwest Indian College in Bellingham, Washington, makes clear that to the Navajo, the ma'ii, Coyote, sometimes also called "Little Trotter," the Roamer, or "First Scolder," was part of the Holy People, formed along with First Man and First Woman. Coyote is a companion to First Man, and often plays the part of scout.

But Coyote isn't honorable or serious. He's constantly horny, tricky, and thieving. His quirks have permanently impacted humankind. When he stole Water Monster's baby, her outrage made waters rise in the worlds below and drove humans up into the world we live in today. He snagged Black God's bag of stars as the god was setting up the constellations and blew stars randomly across the sky. It's Coyote who decided that we will have death as well as birth—and then used his own fur to give humans pubic hair.

Because Coyote decided in favor of our eventual demise, present-day coyotes are also not a great omen. If a coyote crosses the road in front of their car, some Navajo will pull over, waiting until another car has passed to continue their journey, explains Harry Walters, the anthropologist and Diné elder. "We avoid a coyote," he says. "He represents a necessary evil." Because of Coyote's decision that brought death into the world, Walters says coyotes remind the Diné of all the difficult, painful decisions we have to make—the hard choices we make in war or in the hospital. Decisions that might be right, but don't make anyone happy.

In addition to his lust and thievery, Coyote also brags more than any egomaniac you've ever come across. Most of his boasting is in Coyote Stories, in which he features primarily as a trickster. He's not a very good one. "The Coyote in Navajo is someone that cannot leave well enough alone," Walters says. Coyote often boasts that he's the best and tries things only to fall flat on his face. For example, when Coyote visits Wolf, Wolf buries two arrows in the ashes of the fire and pulls out two delicious pieces of meat. When Wolf visits Coyote, Coyote's pretty sure he can do it too. He gets only burned-up wood for his trouble.

Coyote Stories in general are stories of hubris. They, like coyotes to this day, are a reminder that none of us are as big and bad as we think we are. Coyote the character is a stand-in for anyone who has ever gotten a little too high on themselves.

Lately, we've been forced to look into the mirror coyotes hold up for us, as those same animals come trotting in our direction. Over the past century, coyotes have expanded out of the range we generally think of as "theirs"—west of the Mississippi River, south through Mexico, north through the prairie provinces in Canada—to stroll the beaches of Miami and check out Central Park.

The popular version of why this happened is that colonists wiped out wolves east of the Mississippi. Nature abhors a vacuum, and coyotes swaggered into the void. But really, that's only partially true, explains Roland Kays. After all, west of the Mississippi, the presence of wolves never actually kept coyotes down. The spaces they occupy—wolves pursuing larger game, and coyotes pouncing on squirrels, mice, and voles—mean they can live around each other without much trouble.

Yes, killing off wolves would certainly help coyotes cover more area (a wolf will absolutely kill new arriving coyotes who don't have a pack for protection). But humans did more than just kill off one canid for another. "If you look at their original distribution, [coyotes] basically weren't in any densely forested areas," Kays says. "They weren't in the eastern forest. They weren't in the great northern forests. They weren't in the Pacific Northwest rainforest, and they weren't in the tropical rainforest." Coyotes, in short, aren't deep forest creatures. But luckily for coyotes, colonists aren't either. "We have cut down much of those forests . . . which makes it more suitable habitat for them."

Coyotes strode into these new areas at speed. Every generation of coyotes has some members that strike out in search of adventure. "They will go and keep going," says Bridgett vonHoldt, an evolutionary geneticist at Princeton University. "And then they'll have a litter and those pups will grow up and keep moving. So they really can disperse their genes across large geographic stretches."

Of course, the first arrivals to any new area might find dating opportunities sparse. So the waves of coyotes that have spread east have ended up looking slightly different from their western cousins as they settled for a wolf here and a domestic dog there. They're bigger, a little wolfier, a little doggier. VonHoldt and Kays have compared the genes of coyotes in the East to those in the West. "In the northeast, it was on average, eight percent wolf, eight percent dog, and eighty-four percent coyote," Kays says. "The first coyote to enter whatever state or county didn't have any other coyotes to mate with, but there would have been dogs running around. So that's when we think they hybridized. And then as more coyote showed up, they backcrossed into the coyote population."

The results of those first matings linger on, in larger bodies, larger skulls, and different coat colors. "There's a lot of random, weird-looking animals that I've been sent pictures or some DNA of," vonHoldt says. "There was one that looked like a wire-haired coyote." Another was a coyote that could have passed for a German shepherd. I personally am holding out hope for the first coyote-dachshund hybrid to take the internet by storm.

VonHoldt gets the DNA because people kill a lot of coyotes. In *Coyote America*, Flores estimates that five hundred thousand coyotes are killed every year, but there is no federal database keeping track, so it's hard to know. It seems no matter how many coyotes people kill off, there are more to take their place. VonHoldt has spent a lot of time among trappers and hunters, obtaining coyote samples, and one thing comes clear. "They know that as they hunt more coyote, they get more coyote."

Coyotes are like a gas—they'll expand to fill any available space. Kill off coyotes, and the remaining adults may start having slightly larger litters, while younger ones might start breeding slightly earlier. After all, kill off some coyotes, and there's more food left for the rest. "Statistically, the numbers are like half a pup more," vonHoldt says. The increase is so small it's only really a trend. "But when you add that up over twenty, one hundred coyotes, that starts providing more animals." The big population increases, however, aren't from breeding. Instead they seem to come from nearby. In a study looking at pretrapping and posttrapping coyote numbers in South Carolina, scientists found that the remaining coyotes bred slightly more, but the whole coyote population also ended up slightly younger. A few more were being born, but even more were moving in.

In expanding to fill the available space, coyotes often trot into areas humans have laid claim to, and we humans feel unsettled when we see something we don't expect. "In an urban area, we expect companion animals," says Shelley Alexander, who studies human dimensions of wildlife at the University of Calgary in Alberta. "In the rural fringes, we expect livestock, and then in the hinterland, in the third area, we expect wildlife." We often forget that animals don't make these distinctions. "If they go in the wrong zone, they are transgressing."

Alexander ran a set of in-depth interviews to find out how

people viewed the coyotes that lived among them in an agricultural community surrounding Calgary. "I expected them to be much more uniform," she says. Instead, every person seemed to have their own relationship with the environment they lived in.

A few residents—particularly nonfarming residents—saw themselves sharing a landscape with coyotes as equals, and caring for coyotes as part of their identity. Others, though, saw themselves as the dominant animal, and that affected how they viewed other animals on their property. "Some agricultural people will view themselves as visitors on the land, using the natural land for their purposes," Alexander explains. "The other group of livestock owners . . . they own that property. Animals come and go on there . . . it's human property first, animal second." Even when the land they live on was viewed as their property, residents had different ideas of how to treat coyotes on it. They might never kill coyotes. Or they might think they need to cull them regularly to maintain ecosystem balance. And in some cases, animals come such a distant second that people will kill for the fun of it.

In Alexander's study, everyone's tolerance had a limit. There was always a point where people would say a coyote had gotten too close. Maybe that was the porch, maybe it was the yard. One person's measure of "too close" was a couple of hundred meters, she recalls. "His measure was because that's how far his rifle could take the coyote out."

The boundary was one between a place that was human (the farmyard, the fence), and a place where wild animals were "allowed." Coyotes might be permitted to walk a fine line there, but they were never allowed to act—killing cats, dogs, or livestock would always be trespassing. "The idea is that the coyote should somehow be able to consciously understand the difference between what's right and wrong in terms of the location," Alexander notes. But since some residents were coyote lovers, handing out dog food to frequent visitors, while others would shoot anything furry, it's a testament to the coyotes' cleverness that they could figure out the difference at all.

Coyotes didn't just represent wilderness trespassing on human-dominated order. Calgary had well over a million people in 2014, and urban dwellers increasingly moved to rural areas nearby to build a life in a more natural landscape. They—and their beliefs about coyotes—came into contact with the rural population. Some of the agricultural people in the area, Alexander says, "saw those urban people come out and they bring their urban sentimentality with them." That agricultural population felt that those urban "animal lovers" then tried to layer their values "on top of agricultural people, they try to change the laws, they tell them they can't kill things, they report them for killing stuff."

The animals came to represent "the perceived power struggle between urban and rural." Some in the agricultural community felt the urban groups moving out to the country didn't understand the agricultural point of view. "They'll say that urban people . . . don't know what it's like to put a hard day's work in," she explains. "And if they knew what it was like to put a hard day's work into growing crop or caring for cattle, then they would never behave this way."

The power struggle took on a violent edge. One part of Alberta held a coyote-killing contest, and Alexander was asked to speak about it to the media in 2015. "I tried to walk a path in between, where I didn't say anybody was right or wrong," she says. But she also encouraged them to try other control options, because, as vonHoldt says, hunt more coyote, you get more coyote. Those new coyotes don't know the area and might be "more prone to depredate on livestock and others," Alexander says. "So the idea itself doesn't work."

It didn't fly. Alexander had walked into the power struggle. She became the symbol of the urban woman in an ivory tower, telling people who lived on the land what to do. On the other side of the divide, the hunt's organizer said he received death threats. Coyotes, of course, paid the real price. "They put on heavier pressure and they killed twice, at least twice as many," she says, and she thinks she knows why. "To pay me back. To pay us all back."

COYOTES MIGHT TAKE well to the country, and even the suburbs, but you'd think they would leave dense downtowns alone. Wildlife in general doesn't respond well to urbanization, explains Chris Schell, an urban ecologist at the University of California, Berkeley. The closer to downtown you get, the fewer coyotes you will see, compared to the outskirts.

But coyotes can get used to a lot, and "urbanization," it turns out, is relative. All it takes is a little green space. Think of Chicago, Schell says, a densely packed city with a long history. "The history of the city, and the way in which it's shaped, has been that way for long enough that animals that are coming into the city are going to be used to those parameters."

There are good things about being an urban coyote, after all. "The survival rate is quite a bit higher in the urban areas than in rural settings, especially in most parts of the U.S., where hunting and trapping is their largest cause of mortality," explains Stan Gehrt, a wildlife ecologist at Ohio State University in Columbus. Even in the city, though, "we're still the main cause of mortality for them." Then, we kill them with cars, though we don't manage to kill as many. "We're not nearly as effective at killing them with cars as we are with guns and traps."

Gehrt has been studying coyotes in downtown Chicago for nearly two decades and has found that coyotes fall into three categories. The ex-urban coyotes live the farthest from town, in the counties surrounding the city, but unlike the people they live with, they never have to commute. Those areas have big enough parks, forest fragments, and nature preserves, Gehrt says, that coyotes never have to see people if they don't want to. Another set of coyotes is the suburban group. Their environment is patchier. They have golf courses and forests and parks, but they are fewer and farther apart. "These animals have to move among people," Gehrt says. "But there's refuges for them."

The most urban group of coyotes in Chicago is very urban indeed. In maps from Gehrt's paper, collared coyotes can be traced by the lakefront, along and around highways and railroad tracks, right in the middle of downtown. "What little green space is available is usually long and linear, because of the way we develop our travel lanes," he says. "They have to deal with the greatest amounts of traffic and greatest amounts of people."

They can. And they do. "We know litter sizes and the health and physical size of the pups," Gehrt says. There's no difference in health at all. "If anything, they're larger in the most urban areas."

You might think, among the dirt and grime of downtown, or the neat orderliness of suburbs, it would be inspiring to see this shaggy sign of wilderness. Don't people move into the suburbs, after all, to get a little closer to nature? In a survey of people's posts about coyotes on the local social media site Nextdoor, Chris Kelty, an anthropologist and historian at the University of California, Los Angeles, has shown that they certainly do not. Kelty and his colleagues gathered 461 Nextdoor posts from the Los Angeles area about coyotes and examined the posts and any comment threads.

The most common post was a warning, some variation of "I saw a coyote here, keep your pets inside!" It's only one sentence, but Kelty and his colleagues found a lot of meaning. It alerts people to the place (often by providing a cross street) and suggests that the animal doesn't belong there. The warning about pets is a good thing for coexistence—certainly people should keep their pets inside—but it also serves to make people afraid.

The fear and the idea of coyotes as out of place are deeply connected, Kelty says. The coyotes have been living in Southern

California forever, but have fit themselves into a human-created niche—eating fruit from gardens, rats, and sometimes cats. "We've created an urban ecology, and each of these animals is finding a niche. It shouldn't really surprise anybody," he says. "But when you start from this assumption that the city is for humans and wild animals exist elsewhere, then it comes as a surprise to people that these things might be happening."

Schell, who is Black, says he sees a lot of parallels between being a coyote in an urban landscape and being a Black man in America. It's "that experience of being hyperpolarizing," he says. "Doing everything you can to survive in a world that was never really built for you and making it work but constantly being vilified for a whole bunch of other reasons, sometimes completely unfairly." When people's chicken coops get raided, for example, coyotes get blamed, even though it's more likely to be a raccoon, he notes.

That fear and that sense of something out of place are deeply tied to a sense of domination and control, Schell explains. He ran a study of people's attitudes toward coyotes and raccoons in Tacoma, Washington. "The folks that have the most problems with coyotes being in the backyards are those that have more economic standing or those that have outdoor cats," he says. "They believe that coyotes have the right to be out in the wild, just not in their wild. It's superbly colonialist." Coyotes end up vilified, Schell says, even when they do nothing, just for being where they are.

WE CALL COYOTES and other carnivores pests in part because they make us feel vulnerable—fearing for our pets and our livestock, but even for ourselves, when reports of coyote attacks make the news. They challenge our sense of control, revealing that human power over the landscape is an illusion. They show we are weaker than we seem. Coexistence with coyotes, then, depends on helping people feel less vulnerable, giving them back their sense of control.

This can be challenging when dealing with ranching communities who face coyotes—and now, sometimes, wolves as well.
There, loss of control looks like the loss of calves, sheep, and goats.
Being a rancher does mean facing life and death pretty regularly.
But it still doesn't lessen the impact of seeing an animal you cared
for dead in a pool of blood. It's an emotional loss and a real loss of
income.

No one wants to feel helpless, to feel unable to keep the animals
they care for safe. For farmers and ranchers in the United States,
wresting control back has often happened with guns and traps. But
with wolves as a protected species in many areas, and with people
from urban areas urging coyote tolerance, "the rancher groups that
we talked to felt like their hands were a little tied on what they
could or couldn't do," says Julie Young, a wildlife biologist and former field station leader at the National Wildlife Research Center
Predator Ecology and Behavior Project.

As the field station leader, Young didn't just have to manage
people. She managed one hundred wild, captive coyotes in large
outdoor enclosures. Scientists come from all over the country to
conduct behavioral research on the coyotes. There, Young and her
colleagues have been working on ways to help ranchers feel like
they are doing something effective. And they've been doing it with
flags.

Every species has its quirks. For canids like wolves and coyotes, one of those quirks is a weird distaste for flags. In Europe,
Young explains, hunters would set up lines with bits of fabric waving off them—a technique called fladry. "This fladry would create a barrier," Young says, that wolves just won't cross. Wolves are
naturally wary of new, unfamiliar things, and the way flags wave in
the breeze can startle them. Hunters used it to "funnel them into a
corner and then kill them off."

Starting in 2000, scientists transplanted the Old World flag
hunting techniques to the western United States and turned them

inside out. Instead of keeping wolves in, fluttering strips of plastic, like those things you tie around your waist in flag football, keep wolves out of pastures.

When Young later began to spread the word for fladry in the agricultural community, she says, the technique faced some skepticism. "One guy said, 'Oh, no, my place is gonna look like a used car lot,'" she recalls. But hey, if wolves hate used car lots? Look like one. Maybe add one of those wacky waving inflatable people for good measure.

Wolves avoided the flags, coyotes less so. "We started seeing coyotes would cross more frequently than wolves," Young says. "They test, test, test, and then they dart through. And what we realized is, fladry was made for wolves. Wolves are bigger." Young and her colleagues narrowed the gap between the flags, until it looked too small for a coyote to dart through. Success.

But coyotes and wolves are wise. No matter their distaste for garish decor, they will eventually get hungry enough and smart enough to learn the flags don't bite, and there's meat on the other side. So John Shivik, the author of *The Predator Paradox*, has experimented with turbo-fladry. The wire connecting the flags is turbo-charged for an unpleasant zap. "So we can go out on the landscape and put fladry up and help people protect their livestock, and they feel a lot less vulnerable," Young says. "We didn't get losses to zero, but we got them lower, and people felt like they were doing something."

Coyotes might, over time, get used to flags, and electric systems break down. But one thing that never breaks down? Other predators.

JENNIFER WARD, THE natural resources coordinator for Benton County, Oregon, is driving me out to Red Bird Acres outside of Corvallis. In early December, the Willamette Valley glows. Skeins of geese wheel overhead, the air manages to be both misty and crisp at the same time, and even the fields that have been sprayed with herbicide to kill off the cover crop manage to look a bucolic shade of gold. There are small hobby farms, larger, not-hobby farms, exurbs, and a lot of people who want to live close to nature. "What is wildlife habitat and what is people habitat are constantly bumping up against each other," Ward says.

We are met at the gate of the farm by Randy Comeleo, who, with his wife, Pam Comeleo, founded the Agriculture and Wildlife Protection Program in Benton County in their spare time. Benton County lies up against the Coast Range to the west—an area densely populated with cougars and coyotes. Based on the numbers trapped every year, coyotes cause most of the sheep farmers' losses in the area, Randy notes.

The Comeleos didn't really think much on what people did about those losses, until 2011, when someone who lived next to the Oregon State University College of Agricultural Sciences in Corvallis was walking her property line. "She found a dead fawn that had been trapped," Pam says. Later, "she found a raccoon, a live raccoon, still hanging from a snare on the fence." The encounters made the paper, and the Comeleos read it and realized that the county and the university were hiring trappers to try to stop predators getting to their livestock.

"This is the first we knew that Wildlife Services was even working in our county," Pam says. "We felt stupid later when we thought about it, of course they are. But at the time, we thought Wildlife Services was out there, you know." Out there, in the wilderness. Predators, and their trappers, weren't supposed to be quite so close to home.

Public pressure made OSU deactivate the snares in 2012, but the university has since kept its contract with the trapper provided by the USDA. So the Comeleos decided to try the county route to reduce wildlife trapping. Inspired by a similar program run by the Marin County Department of Agriculture in California, they proposed a grant program to the county administrators in 2017. Farmers could apply for grants of up to five thousand dollars to buy things that would keep predators away, without trapping or killing them. The county would also maintain what is perhaps the strangest little free library—a library of nonlethal predator tools, like Critter Gitters (motion-sensing lights and alarms) and iridescent fladry (good for deterring birds). That way farmers can try before they buy.

The county agreed, and the Agriculture and Wildlife Protection Program started taking grant applications in 2017. Though the Comeleos founded the project, and Randy still reviews grant applications and provides a lot of input, by the time of my visit, Ward is running the day-to-day operation, a role the county balances with paying for the trapper through the USDA.

There were some objections at first. Some farmers, Pam Comeleo says, "don't want the lethal program to go away. They're just entrenched somehow in that mindset. . . . We met people who just hate coyotes, I mean literally hate them."

The first takers on the grant were not those people. Instead, a lot of the applicants never wanted to use lethal methods, or were willing to try something different.

First-generation farmers Laura and Robin Sage, who own Red Bird Acres, pride themselves on raising livestock according to strict animal welfare guidelines and never want to use lethal methods. When Ward and I arrive at their farm, Robin Sage greets us, and we trudge down the hill across clumps of long, wet grass into a series of wire-fenced pastures—part of his leased eighty acres. Some house a few chickens, midmolt with comical bare necks, looking like they're dressing up as vultures for Halloween. Others house free-range pigs and a joyful wriggle of piglets. A couple of goats blend in with the pigs. The rest of the acreage had housed turkeys for the season, but it's after Thanksgiving and they are now working their way through Oregonian digestive tracts.

As we walk down the lane, though, what stand out most are the large lumps of white hurtling toward us. Four huge Maremma Sheepdogs come romping over to be told that they are very good dogs indeed. Between ear scratches and belly rubs, I marvel at just how gigantic they are. When one or two of them leap a little, it's clear that they could easily rest their front paws on my shoulders.

This pack—dignified Shasta Lion (technically a Maremma/Pyrenees), cuddlebug Lassen, and frat boys Atreyu and Falkor—is Sage's antipredator defense system. The dogs live outdoors in the pastures with their charges, coming in only if wildfire smoke gets dangerous or the heat too strong. They have thick, luxuriant coats that make winters no big deal. This species, with the right training, will defend almost anything (Maremmas are famous in Australia for defending the little penguins of Middle Island from foxes).

When I ask if he has coyote problems, Sage gestures to the dogs. "Not anymore." Before the dogs, though, "we could be here and you just hear a cacophony of coyotes just echo around the property." Now, they're mostly gone, and Sage is confident the dogs are the reason.

He's also got electrified fencing at the outer edge of pastures, mostly to keep his stock in rather than other things out. The dogs take care of the latter. "It was a constant struggle with mostly aerial predators, like owls. And hawks were like the main thing with the chickens," Sage says. The Maremmas keep out the coyotes, skunks, raccoons, foxes, hawks, and more that might see Sage's animals and think of a free meal. "They'll just go over and bark at things," he says.

The Sages have had the stately and responsible Shasta since 2015. But they quickly realized one dog couldn't do it all. Lassen, then, is the result of the Comeleos' Agriculture and Wildlife Protection Program grant. The Sages applied for a grant to purchase the dog (puppies go for about eight hundred dollars, Sage says), electrified fencing, and some strobe lights that help keep owls and other predators at bay. After that, Sage knew he needed more, and Atreyu and Falkor joined the pack.

It has made a difference. In 2015, Red Bird Acres lost more than 150 turkeys and chickens; in 2017, with Shasta's help, they lost five. In 2018, they had lost just one. Sage continues to lose a few fowl to owls, but at this point, he says, "it's a donation to the Red Bird Acres owl fund." He recognizes that nature will take some of its due. The dogs, fencing, and lights keep it to a minimum.

Strobe lights might not seem like an effective deterrent against a 140-pound predator. And yet, that's what Scottie Jones has protecting her sheep at Leaping Lamb Farm over on the other side of the Coast Range. Her farm—sixty-seven acres that she and her husband have owned since 2003—is in a steep-sided golden cup of a valley with a salmon-spawning stream at the bottom. "A predator highway," Ward says, as we tour the property.

Jones also applied for the Agriculture and Wildlife Protection Program grant, but her money went to portable electric fences and flashing red strobe lights. She now has more than forty lights strung around her property, protecting sheep, goats, chickens, ducks, a peacock named Elton, and an ancient, gigantic turkey named Gandalf. Her three barn cats—one, Lucy Goosey, accompanies us on our tour—help with the mice, rats, and moles.

The final portion of her antipredator defense is more prosaic—an old boom box, playing the radio out over the pasture all night. What it plays "depends on what station we can get," Jones says. "It was rap for a while." The goal with the lights and noise is to make it sound like a human is nearby.

Before the lights and sound, Jones was losing lambs. One year, she lost nine. It was a cougar, sliding through the steep, Douglas fir–lined sides of the valley and leaping effortlessly over her fences with young mutton in its jaws. She called the county trapper, but he was never able to catch anything. She also had raccoons trying to get to her poultry. "We came out and shot them out of the trees," she says. "Me in my nightgown holding the flashlight and my husband shooting."

Jones applied for the grant because "it was there." In addition to the lights and fencing, she brings her animals in at night, putting the sheep in pastures closer to the barn and farm-stay house. She tried keeping them out in the pasture with the lights, but "it's too hard on my heart," she says. She was too afraid of losing another sheep.

Her changed behavior and the night-lights have completely eliminated her cougar problem—even when the neighbor two doors down lost goats, she never did. She's still losing some chicks to weasels, but is strengthening her pens.

Farmers can just call the trapper, of course. His services are free. But they would have to call at least once per year, and they'd lose stock every year too. Nonlethal management requires more work and more money. And, Ward says, it requires a different mindset. These deterrence methods can't just be set and forgotten, she says. "You have to continually be involved with your animals, the deterrents that you're using, right? If you're willing to do that, they're quite effective."

The mindset is one of coexistence, one that recognizes that maybe we shouldn't have total power over the landscape we live in. One that accepts that we need to make long-term, proactive changes, not just retaliate when things get bad.

SCIENTISTS ARE ALWAYS looking for new ways to keep coyotes away—both from livestock and from suburban pets and humans. When Niamh Quinn hacked off that dead coyote's head, she wasn't playing out some sort of gruesome murder fantasy. The heads were bound for the lab of Ted Stankowich, who studies mesopredators at California State University, Long Beach.

When I arrive in the lab, the smell knocks me back a pace. It's not the reek of a decomposing necropsy. Stankowich's lab studies skunks as well as coyotes. The entire lab is permeated with the stench.

Stankowich and his student, Lizbeth Jardon, assure me that "it will go away in a minute." An hour later, it is clear that they are wrong. Either working in the lab for years has rendered them insensible to eau de skunk, or they are having fun with the journalist.

Jardon is the student who gets the coyote heads. She expertly slices off jaw muscles and boils the bones clean, then creates a careful 3-D model of each skull. Using the dimensions of the muscles and the skulls, she's trying to calculate the bite force of the coyotes of Southern California. She and Stankowich are hoping to find out how the food opportunities of urban and rural coyotes might affect which animals are successful in which environments. For example, Jardon says, a coyote in an urban environment might not be hunting as much; it might get more fruit, trash, and pet food. Over time, the urban coyote and its descendants might get away with weaker jaw muscles. The rural coyotes, on the other hand, might live a harder life and have harder jaw muscles to prove it. They've found no differences so far, but hope and new coyote heads are constant.

Weak jaw muscles can still worry pet owners. As I walk into the lab, Stankowich gestures to an odd contraption by the door. It's a bright yellow Kevlar-reinforced dog coat, outfitted with side rows and a neck collar of pointy metal spikes and topped with neon orange bristles. It's exactly what you'd put on your miniature poodle if you wanted it to look both very punk rock and completely ridiculous. It's a CoyoteVest (as seen on *Shark Tank!*), and it's meant to make your tiny, walking, barking coyote snack look like a bad idea. The maker dropped by to get Stankowich's ideas and left him with a sample. "It's an interesting product," Stankowich says. Does it work? "[The maker] says it does. He doesn't have any data on it, and there's really no way to prove that it works."

Other students of Stankowich's lab, like Kathy Vo and Caitlin

Fay, are the main reason the lab stinks. They have both studied the one thing a coyote will regret eating—the striped skunk (*Mephitis mephitis*). "I actually have a poor sense of smell," Vo says. "So it was like I was meant to work in this lab."

Fay began their investigation, spending a summer with Julie Young's captive coyotes in Utah. She cut chunks of plywood into flat, long hexagons and covered them in fake fur. She also attached a sprayer underneath, full of diluted "skunk anal gland secretions," and put some tasty bait in the center of each furry plate.

The trick was that the fake fur was of different colors. Brown coated "prey" didn't spray. Striped black-and-white prey did. The coyotes weren't that thrilled with the skunk-striped models to begin with, and it only took about two sprays for the average coyote to figure it out. They then avoided plates with black-and-white striped fur—and plates with black-and-white blocks and V shapes. Some even avoided spirals, for good measure.

It's an example of classical conditioning, teaching coyotes to associate a stimulus—skunk stripes—with something very bad—smelling like a skunk. Skunks, scientists hypothesize, have the bold stripes they do as a warning signal. Try to eat me, and you're gonna get it.

Vo has since spread fake skunk models into real skunk habitat in California, coupled with motion-capture cameras, to see how the coyotes react to high contrast black-and-white stripes, spots, or spirals, compared with gray or brown. But of course, Vo notes, other animals see the fake skunks too. Including other skunks. "Another part of my thesis study is looking at how skunks react to the fake skunk models," they explain. "We surprisingly saw real skunks really digging my models." When a skunk meets another striped animal in the moonlight, romance can bloom, especially if the object of your affections doesn't run away. "We did observe a decent amount of . . . mounting behavior, as I would call it."

But Vo is also looking at how high-contrast coat colors might keep coyotes away from other animals, like dogs and cats. They sent out an online survey asking for descriptions of people's pets, their size and breed and how much time they spent outdoors every day. Asking people to talk about their pets is an easy sell. About 1,300 dogs and 590 cats later, Vo showed that cat color didn't matter. What mattered most? How much time they were spending outside. Both dogs and cats that spent time outside experienced more negative interactions with coyotes than pets that didn't. Smaller dogs were the most likely to end up on the business end of a coyote since a coyote is far more likely to consider a smaller animal dinner. But dogs with high color contrast—lots of black and white—did have less severe run-ins with coyotes.

Young has worked on ways to keep more urban coyotes away from people as well as pets. Hazing—the process of making a ton of noise, flashing lights, making a big commotion—is what people are told to do when they encounter a coyote. But coyotes have long memories and encounter many people in their lives. Some people will haze, some will take photos, and some might even proffer snacks. All of these affect how a coyote will react when it sees a human the next time.

In a small study using herself, pairs of captive coyotes, two nine-to-ten-year-old kids (with very chill parents), and a dog (a black Lab named Iris), Young showed that hazing did, overall, work. If you yell and shake cans of pennies and stomp your feet enough, a coyote will move off. But if coyotes have prior experience with humans offering food, they'll stick around a lot longer. Walking the dog nearby also required more hazing effort, Young says. The coyotes were way too interested in the other canid to pay attention to the yelling, penny-shaking human.

It's even harder if the people doing the hazing are kids. "The coyotes really never saw the children as a threat, even when they were hazing," she says. To make the coyote move, the kids had to work a lot harder.

Like the CoyoteVest, though, a good hazing isn't really about what the coyote does. It's about how vulnerable people feel— coyotes can be dangerous, in the way any medium-large canid can be. In 2009, a Canadian folk singer named Taylor Mitchell went for a hike in Cape Breton Highlands National Park while she was on tour, and got attacked by a pair of coyotes. Two other hikers found her and got help. Mitchell was transported out of the national park and eventually made it to a hospital in Halifax, but she died there of her injuries. She was nineteen. Mitchell remains the only adult killed by coyotes ever recorded.

Cape Breton is an island that only got a causeway to the mainland in the 1950s, and coyotes were a recent arrival. The attack was a shock to the small community that lives there, says Carly Sponarski, a conservation social scientist with the Canadian Forest Service in Alberta. "It was traumatic," she says. "Lots of the community suffered from post-traumatic stress from that one instance." In such a small community, it seemed like everyone knew someone who had encountered Mitchell that fateful day.

Taking surveys of visitors, park staff, and residents near the national park after the attack, Sponarski found that residents were far more afraid of coyotes than the park staff or tourists were. The residents felt they had almost no control if a coyote approached them, and rated the risk of seeing a coyote as much higher than it really was. "They felt like the coyote took their backyard away," Sponarski says. The beautiful park where they loved to walk, which backed onto many of their properties, had become dark and dangerous.

"Driving our car every day to work, [we have a] much better chance of being hit and having a bad accident driving than we ever have just going out into the forest once in a while," she notes. "Because we live with fear, we get used to it." Over time, we become desensitized to something like driving, which is very high risk. But something low risk—walking in the woods where coyotes are present—can feel far scarier than it is, because it is so very rare.

The one-time traumatic experience that Cape Breton residents suffered left them feeling powerless in part because it did happen only once. Normal, nonlethal interactions with the coyotes didn't occur every day. "And so that risk, that perceived risk will never be moderated to the appropriate level," Sponarski says.

She decided to give the residents their power back. She set up a series of educational activities. First, residents got to talk over their experiences with coyotes and how they felt. Then, they ranked their perceived risk of things like driving, hiking, and running into coyotes, and talked over the real risk of those things.

Then, the residents got to look through the eyes of a coyote. Examining maps with GPS-tagged coyote locations, they worked to determine what a coyote's day was like. Sponarski and her colleagues talked through typical coyote behaviors—including the yips and howls that had haunted the residents. "It's not that much different than someone saying, at the end of the day, when we come home, 'Hey, honey, how you doing?' and giving you a big hug," she explains. "You know, they're using these vocalizations to reconnect, it's dinner conversation in a way."

The residents also got to practice coyote defense. Using a map of a typical Cape Breton yard, Sponarski had them play "pin the coyote," putting stickers on coyote attractants. Residents learned from each other what attractants might be and how to get rid of them.

Finally, she helped them with their biggest fear—the loss of control they felt over whether they might see a coyote and what might happen if they did. She taught them how to haze. "We took what [the Cape Breton Highlands National Park] was saying," she says. "Which was BAM: back away, act big, make noise, and we added an F to it." The F stands for "fight back."

She had the residents practice their defensive moves, using a wooden coyote on wheels named Chum. Sponarski handed the residents sticks, and then sicced Chum on them, rolling him forward on his wheels. "Women were vicious," she says. "Poor Chum, he lost more ears. He, like, got pieces of his nose taken off . . . they appropriately went for the head, which is what you want to do in this situation."

As the residents studied GPS signals, spotted coyote attractors, and attacked Chum, she watched them change. "They felt more empowered, and less vulnerable," she observed. Sure enough, when surveyed after the program, residents were less afraid. Knowledge was power.

In their survey of people's posts about coyotes on Nextdoor, Chris Kelty found that it wasn't whether a coyote had actually done anything that mattered. What mattered was what the knowledge of the coyote's existence did to people's sense of power. "The thing that probably unites the anticoyote people in that space is . . . a desire to live without any fear, to minimize the risk almost entirely," says Chase Niesner, a graduate student studying under Kelty, who performed interviews related to the work. People imagine "what it means to live a life worth living. That is punctured by the presence of the coyote, whether or not the coyote is actually eating their pets or not. Just the presence of the coyote makes these people less able to enjoy their ideal life."

Niesner and I sat together in Griffith Park, absorbing the late afternoon sun and wondering what life could be like if we could acknowledge that our environments—even our cities and suburbs—will always carry risks. If we could accept our lack of power and meet it with preparedness, instead of preemptive strikes. If we could loosen our fearful grip on our landscapes, just a little. There is risk, but there is reward too, for the people and the animals that live there.

As we talk, I hear a rustling to my left. A coyote trots into view on the hillside. It comes toward us and pauses, watching and waiting. Its posture is relaxed—it's at home here. After a few seconds, it recognizes that we have no food and trots over to a man waiting in the shade a dozen yards away—a man who appears to be waiting for it, holding something in his hands. It could be food. Another coyote follows, and a third moves through the brush farther up the hillside, the sound and shadow suggesting that this one has only three legs.

On the sidewalk a woman walks a small dog. They don't seem to see the coyotes at all.

I was surprised to see them—it was still broad daylight, not dusk—but I wasn't afraid. This park was shared territory. I knew we wouldn't be alone, and with my knowledge, I did not feel powerless.

8

A FLUTTER OF SPARROWS

n December 1958, then ten-year-old Mingli Sun was stabbing rodents with the spindle of a spinning wheel. He could kill more than ten in a night. Sun was in the fourth team of the Zhongyuan Production Brigade (under Communism, instead of villages and towns, people were divided into production brigades and teams), in the Zhonghao Commune, in Hubei Province, China, and he was doing his homework.

Sun was playing a part in the Four Pests campaign. Created by Mao Zedong, it was part of a set of sweeping public health campaigns that took place throughout the 1950s, hitting their peak during the Great Leap Forward. The goal was to bring China from its feudal, agrarian economic past to a bright Communist future. As part of it, Mao wished to exterminate four "blights" on the people. Two of the pests were a target because they carried disease: rodents (mice and rats) and mosquitoes. Another was a well-known nuisance—the housefly. The fourth ate seed grain, endangering the future of China's crops. The Eurasian tree sparrow (*Passer montanus*) had to go.

In a 2012 paper on the history of the sparrow elimination campaign, Weimin Xiong, a historian at the University of Science and Technology of China, suspects that Mao might have had personal reasons to take on sparrows. "For a long time, the majority of Chinese peasants couldn't get enough to eat, and they were extremely poor," a translation of the paper reads. "The sparrow, who would come into their fields and peck at grain and compete with them for food, was widely disliked. As the son of a peasant, Mao most likely hated sparrows from a young age."

The sparrow elimination campaign also had some public image benefits, Xiong writes. People could kill all the flies and mosquitoes they wanted, but eliminating standing water or cleaning up trash to stop insect infestations doesn't look very impressive. "The directive to smash sparrows was the one that could most easily be implemented," he notes. It was also easy to show off your results. With rats, mice, and sparrows, you could make imposing piles with the bodies.

The efforts to eliminate sparrows began in 1955, as a youth movement, but by 1958, everyone had been recruited to help kill the birds. "The school asked us to submit ten mouse/rat tails or ten sparrow legs at the start of the semester," Mingli recalls. Sparrows, however, were tough to catch. The Chinese Communist Party gave out several handbooks with advice. "We tried the method described in *From Hundred-Grass Garden to Three-Taste School*, Lu Xun's prose article." This method involved an upside-down basket with some grain underneath, propped up by a stick. Children would wait until the sparrows hopped under the basket and pull the stick, the classic box trap. Unsurprisingly, "it didn't work."

Sun had an easy target in the rodents in his house. "I lived in a thatched cottage then," he says. "I turned on the flashlight suddenly and saw mice/rats staying on the purlin. They probably assumed that I could not see them." They assumed wrong. Sun attacked and got his homework done. It was, he admits, kind of fun. "People played a cat-and-mouse game with [them]," he says. "Children all enjoyed it then.

"Later, some [people] wanted to eat sparrows, so they used bird netting to trap them, which helped us to accomplish the task assigned by the school. Sometimes we used slingshots to [catch] sparrows. But it was not easy as well." Sun ate sparrow himself. It might seem like there's not good eating on a tiny bird, but they were on menus in some restaurants. He doesn't remember too much of the flavor, but recalls it was "a bit sour."

Sparrow elimination took cities by storm in the form of coordinated mass campaigns. On a spring morning in 1958, people poured into the streets of Peking (now Beijing). Encouraged by loudspeakers blaring about their "cunning and tricky" enemy, the city went to war against the sparrow. They waved red flags, hefted scarecrows hung with bells, and banged on gongs and drums, anything that made a racket. Young boys used shotguns, and girls kept sparrows from resting on wires with long poles. By the end of that spring day, Peking newspapers boasted of between 310,000 and 800,000 sparrows killed.

Mikhail Klochko, a Soviet chemist assigned on a scientific mission to China, was also in Beijing, and wrote, "We Russians watched the slaughter of the sparrows with disgust, and those whose names were Vorobyov [which means "sparrow"] . . . gloomily joked about the mortal danger that threatened them."

"On Sunday, April 20, I was awakened in the early morning by a woman's bloodcurdling screams," wrote Klochko. "I saw that a young woman was running to and fro on the roof of the building next door, frantically waving a bamboo pole with a large sheet tied to it." The guns, noise, and sheets were designed to keep the sparrows from landing or foraging. Villagers shook smaller trees furiously to prevent the birds from resting anywhere. "It was claimed that a sparrow kept in the air for more than four hours was bound to drop from exhaustion."

At first, everyone participated in the sparrow hunt, but screaming, yelling, and shaking trees did not do too much to actually put food on the table. In richer, more urban areas, cities and towns might be able to afford to have everyone take time off from work to kill sparrows, explains Miriam Gross, a Chinese historian at the University of Oklahoma. But "if it's a very poor area, what it may do is delegate this party work to those who are not full producers." Any day you're spending killing pests is a day you're not spending in the fields. After a while, most people went back to work, leaving the pest project to children like Sun, and the elderly.

People posted sentries, smashed eggs, and ripped nests apart. The Eurasian tree sparrow went nearly extinct in China. Overall, Shanghai reported killing 48,695 kilograms of flies, 930,486 rats, 1,213 kilograms of cockroaches, and 1,367,440 sparrows. The final statistics are impossible to rely on, explains Fa-Ti Fan, who studies science in modern East Asia at Binghamton University in New York. No one was really keeping numbers that accurate. But "according to kind of partial statistics [there were] probably two billion [sparrows] killed in the 1950s."

And then the insects arrived.

It turns out that sparrows did eat grain, yes. But when they are raising chicks, they get massive protein cravings, which they satisfy with bugs. In the years after the Four Pests campaign, farmers began losing more of their crops to insects. In Judith Shapiro's *Mao's War Against Nature: Politics and the Environment in Revolutionary China*, an agricultural chemist recalls "in 1959 there were more insects. It wasn't something you could notice for yourself, but our Plant Protection Department measured more infestations in the grain."

Grain output (including rice, wheat, millet, and corn) was certainly suffering. In 1958 the yearly grain output was estimated at

200 million metric tons, but people in some areas of China were already hungry. By 1961, it was 137 million metric tons.

The government blamed the crop decline on drought, which certainly played a role. But in a 2021 preprint paper that ran economic models of grain output, Ruigang Bi and his colleagues suggest that sparrow killing could have accounted for 33 percent of the crop reduction between 1958 and 1961. If you assume that a single sparrow eating pests prevents the loss of 3.3 kilograms of grain, and that 2.1 billion sparrows died, and pesticide could save only so much, the result is an annual loss of 7.1 million metric tons of grain. The loss of sparrows, they estimate, cost enough food to feed twenty-eight million people.

In 1960, Mao called off the sparrow hunt, now declaring bedbugs the fourth pest. But the damage was done. The Four Pests campaign probably contributed to what would become the Great Famine, which killed somewhere between fifteen and fifty-five million people, though the true number remains unknown. The human mind can't even process that much suffering. People ate the bark from trees and clay from pits. They ate human corpses, and killed and ate the living.

The Four Pests campaign wasn't even about the pests at all. It was about power. It was coupled with huge projects to flatten mountains, fill in valleys, straighten rivers, and generally turn China into the country that the Communist Party dreamed it could be. It was, in theory at least, about the people finally conquering the one thing that truly held them back—nature itself. But human politics, and the environment, had other plans. When we exert power over what we don't fully understand, the results can come back to bite us.

IN THE EARLY 1930s, farmers in Queensland, Australia, were attempting to build a sugarcane industry. They were plagued with a lot of problems. One of them was and continues to be cane grubs—

the babies of several species of sugarcane beetles that gnaw away at the roots of the cane. Desperate to get rid of the beetles, scientists at the Bureau of Sugar Experiment Stations began casting around for something to eat them.

They came upon new research on the cane toad (originally *Bufo marinus*, now also *Rhinella marina*, with pit stops at twenty other Latin naming attempts, including *Bufo lazarus*, *Bombinator horridus*, and *Chaunus marinus*—taxonomists, get your act together). They are a quintessential brown warty-looking toad that can get up to 1.5 kilograms in mass and twelve to twenty-five centimeters in length. Raquel Dexter had dissected a bunch of toads that had been introduced onto sugar plantations in Puerto Rico. "What she discovered was that they eat a lot, and that quite a few of the things they eat are insects that would be regarded as pests of sugarcane agriculture," explains Rick Shine, an ecologist and evolutionary biologist at Macquarie University in Sydney.

Based on that idea, the scientists thought that cane toads might eat up the cane grub problem. Like finding diseases for rabbits, cane toads were another example of biocontrol—controlling pests with their natural predators or diseases. And it seemed really positive, at first. "The alternative was things like blowing up the soil and, in particular, dispersing very nasty chemicals that proved to be carcinogenic," Shine says. "And in a sense, these guys were pushing an environmentally friendly form of pest control."

But even the greenest pest control needs to be tested. In an ideal world, the scientists would have conducted rigorous experiments. Maybe they would fence off test plots, see if cane toads decreased grub numbers and increased sugarcane yields. They would have looked at toad breeding patterns and potential for spread. They would have checked to see if the toads might be a threat to any of the native fauna.

They did none of these things. Instead, they dispatched entomologist Reginald William Mungomery with a ticket to Hawaii to bring back toads. "Mungomery and crowd don't seem to have really done very much due diligence," Shine says. "I guess there was political pressure. The cane growers were a powerful lobby group." They didn't even make sure the toads didn't have any diseases before they brought them over—toads in Australia carry lungworm to this day. At least, Shine says, they pulled off the ticks first.

Mungomery and 101 cane toads arrived in Queensland. Within short order the toads bred, and thousands more toads hopped onto sugarcane plantations. And the scientists—and farmers—began to realize they had made a terrible mistake. "Showing that toads eat bugs is a long way off showing that they actually influence sugarcane yields," Shine says.

Cane toads will definitely eat cane grubs. They just need to get to them first, and therein lies the challenge. "Toads don't dig down into the ground, and the grubs are underground," Shine says. "What toads do is they will eat the beetles that are coming to the ground to lay their eggs." The beetles also fly. Toads do not. "The amount of time that the beetles would be on the ground was probably too low for the toads to have a big impact," Shine explains.

So the toads didn't solve the grub problem. The real issue, though, didn't turn out to be what cane toads were eating. It was what ate them. Each adult cane toad appears to wear large, timelessly fashionable shoulder pads. When the toad is grabbed and squeezed (say, in a predator's jaws), the pads exude a milky white poison, and a native animal finds it has accidentally eaten its last meal.

Predators such as quolls (adorable catlike marsupials, genus *Dasyurus*), crocodiles, snakes, and goannas (monitor lizards, genus *Varanus*) fell for the cane toad trap. "The research shows that the impact is horrendous," Shine explains. "But it's visited [on] a relatively small number of species." Cane toads haven't driven any species to extinction, but they have made some native predators far more rare. Those predators also ate mice and rats. And once one of the predators ate a toxic toad, they weren't around to eat rodents, which then became more numerous.

The toads also proved their reproductive prowess. They have a love of human-associated landscapes—places with lights at night that attract tasty bugs, and watered lawns that provide places to mate and lay eggs. So people began to see toads. And then they saw more toads.

From Queensland, the cane toad began a surprisingly speedy westward hop across northern Australia, as well as southward toward Sydney and Canberra. No one has yet been able to stop them. (Farmers eventually found effective pesticides against the cane beetle, but they remain a major cause of crop loss. The cane grubs, like the cane toads, persist.) The toads now occupy a large swath of northern Australia, spreading from the east coast westward. And the cane toad tide is still coming, currently hopping through the Kimberley, the northernmost area of Western Australia.

For the history of the cane toad saga, there is nowhere better to start than the 1988 documentary *Cane Toads: An Unnatural History.* I have vivid memories of watching this film at summer camp as a child, and watching again as an adult did not disappoint. It's got horror music and creepily giggling children cuddling giant toads. A naked man is horrified to find toads spying on him in the shower. A cane toad avenger swerves his van back and forth over the road, appearing to flatten cane toads with disturbing popping noises (don't worry, the van was actually squishing potatoes, no cane toads were harmed in the making of this film). An old man in thick glasses notes emotionally how much he loves to watch cane toads mate.

I can't recommend it enough, honestly.

The film was wildly popular in Australia, explains Shine, and helped to cement the cane toad as public enemy number one. "The toads are just a delightful lightning rod to demonstrate the sometimes bizarre reactions that folks have to some new challenge," he says.

It doesn't help, of course, that cane toads are, well, toads. They've got creepy slit pupils and they're big, muddy-colored, and chunky. "People think they're ugly and don't like them, and it's almost a social license to hate them," Shine says. "They have kind of gone up toward the top of the list in the public's perception of horrible invaders that we need to do something about."

While people deliberately ran over toads, trapped them, and whacked them over the head with pipes, nature was doing its own biocontrol. The cane toads, it turns out, aren't as devastating as scientists predicted. Shine's research has shown that there are high concentrations of toads only at the leading edge of the invading wave. After that, the populations calm down. This isn't because the toads have eaten up all the food or anything. Instead it's because they have integrated themselves into the ecosystem.

Some species of native rodents and birds safely chow down on toads with no ill effects. Kites (a small raptor) in the city of Darwin delicately yank out the toads' tongues—getting a snack without getting poisoned. Water rats, by contrast, flip the toad over, cut neatly into the chest, and go to town on the animal's hearts and livers, avoiding the skin and poisonous shoulders. They might have learned to do this, or it might be an adaptation from handling other poisonous species.

In some cases, species are physically adapting to the presence of toads. Red-bellied blacksnakes in Australia at first died from toad toxins. But over time, the species began to avoid eating toads—and also gained some resistance to their poison. The blacksnakes—as well as another species, the Australian tree snake—even started developing smaller heads compared to their body size. A small enough head, after all, can't eat a big toad.

The toads have changed too. They've got tougher skins that lose less water in the dry Australian heat. The cane toads on the leading edge of the invasion wave are capable of spreading farther than toads left behind. And when there's little food around, toads get violent—they show more cannibalism in Australia than they do in their native South American home. "We've got a good distinctive Aussie cane toad that has adapted both to Australia and to the process of invasion," Shine says.

Shine has studied the effects of cane toads on the native fauna of Australia for decades. And he no longer thinks they're public enemy number one. If Australia managed to eradicate them, quolls and goannas might do better, but it would be bad news for other species who benefited when the quoll and goanna populations got smaller. "Toads are part of that food web," he says. "They have very complex effects up and down . . . they're part of the system. We'd be better off without them. But they're not an A-grade ecological catastrophe."

Now, scientists are still trying to manage cane toads, but actually getting rid of them is not really on the table, says Andy Sheppard, ecologist and research director for managing invasive species and diseases at the Commonwealth Scientific and Industrial Research Organisation in Australia. "At the end of the day I think the cane toad has got the upper hand." Instead, they're trapping tadpoles and looking at potential genetic alterations to make the toads less poisonous. They're even feeding small toads to local wildlife—making them sick to give them a chance against the amphibian onslaught.

The Kimberley, Australia, is the current front line in the toad war. It alternates between extremely wet monsoon seasons and extremely dry periods, and is one of the hottest parts of Australia. There, Georgia Ward-Fear, a wildlife biologist at Macquarie University, is trying to teach wildlife to leave the toads alone before the toads even arrive.

The method is called "teacher toads" and takes advantage of a well-known behavior called "conditioned taste aversion." Give an animal something yucky that makes them feel sick, and they learn very quickly that they should avoid anything that looks, smells, or tastes like that yucky thing. Many of us are familiar with this from that one time we had a terrible food poisoning experience, and we still can't stand the smell or taste of tuna, milk, or whatever your personal poison was—even if it's been decades.

Ward-Fear is currently trying to teach that same lesson, using toads instead of tuna. By exposing predators in the area to small, living but less-deadly toads, poisoned toad-flesh sausages, or dead toads laced with nausea-inducing chemicals, she's trying to make sure the animals avoid the real toads when they do appear.

When I talk to Ward-Fear, Kununurra, the town in the Kimberley where she's working, is at the end of the dry season, and she's out to save the freshwater crocodiles. In the wet season, she says, the crocs aren't in danger from cane toads; they are spread out on the landscape and there's plenty of larger prey. In the dry season, though, water bodies disappear. The hungry crocodiles get confined to a few crowded, muddy areas. When the cane toads arrive, they also end up concentrated around those croc-filled pools during the dry season. A starving, confined croc gives in to a deadly temptation. "They become very vulnerable," she says. "And we start seeing mass mortality events." "Mass mortality event" is scientist-speak for a landscape full of bloated, dead crocodiles who never lived to regret their final cane toad meal.

To teach these crocs to avoid the toads, Ward-Fear has spent the dry season tossing chunks of dead toad to crocodiles. "The toads are cut in half," she says, to avoid the toxic shoulder pads. "It's the back half with all of the organs removed because the females' eggs are highly toxic as well." The toad butts aren't poisonous, but they are laced with a chemical called lithium chloride. It's completely harmless; no animal will get poisoned. They will, however, feel the most intense nausea they have ever suffered in their lives. It's a potent message.

Not every predator gets the message the same way. Another lab taught quolls not to taste toads using cane toad sausages laced with thiabendazole, a dewormer that also induces a strong urge to hurl. The goannas, however, don't take dead prey, so poisoned toad butt or toad sausage is unlikely to lure them in. Instead, Ward-Fear releases cane toad tadpoles into the area ahead of the big cane toad wave. The tadpoles develop into small toads—the teacher toads— and goannas eat them. They get sick, but they don't die like they would eating a large, potent adult toad.

Releasing cane toads in an area that cane toads are about to hit is a tough sell, though. "It just seems like madness," Ward-Fear admits. "It's entirely counterintuitive." That's because people don't realize what the invading wave will do. "They can't really fathom the carnage that they're about to see." She had to pull together government entities and big conservation groups to publicly back the initiative. It has been working. With teacher toads, Ward-Fear has found that about half of the goannas learned their lessons well. Without them, only one of thirty-one released lizards made it.

Cane toads are a justly famous tale, and might leave a poor taste associated with words like "biocontrol." But done properly, biocontrol can actually be effective, Sheppard notes.

The problem with toads is that they aren't picky eaters. "Vertebrate animals generally are not specific predators," he says. "So by definition, they're not good biocontrol agents." Toads, along with cats, dogs, rats, and more, will eat basically anything. So if you deploy them to solve the problem of another animal, they may not eat the animal you want them to eat, and they will most certainly eat things you don't want them to eat.

Invertebrates, however, can get a lot more specialized. "If you look at the insects on any plants in your backyard, you know there will be a lot of generalists, like snails and so on," Sheppard says. "But there will always be one or two species that are entirely dependent and have evolved to feed only on that particular species. And those are the ones we're looking for." The picky eaters of the insect world are the ones that are the best options for biocontrol, he says. "It's the same with weeds. We've been using herbivorous insects and other arthropods and diseases to control weeds for the last one hundred twenty years now. The track record is very high."

Mungomery and other scientists who were interested in the cane toad were emboldened in the first place by the amazing success of biocontrol for the prickly pear. The first prickly pear cacti arrived in Sydney Harbor in 1788. Apparently, colonists wanted them because of the cochineal insects on them—which produce the dye that makes the famous British redcoats.

Colonists liked the insects, but the prickly pear liked Australia. Once the spread got out of control, scientists started looking for something to eat it. They tried a few species, and hit on the cactus moth (*Cactoblastis cactorum*). The scientific name held promise, and the moth did too. Between 1926 and 1931, Australians released more than two billion cactus moth eggs. They did indeed devour those cacti. The cactus moth will eat cactus and only cactus, and for biocontrol, that was a superpower. After such a success, it seems like a no-brainer to start looking for other biocontrol agents.

Biocontrol can also mean deploying diseases such as viruses, including the calicivirus that is used to keep the rabbit populations down in Australia, and a herpes virus scientists are developing to combat grass carp. Again, the diseases they choose are specialized—so specialized that they will infect only the things they are supposed to (in theory, at least).

Cane toads taught Australia a valuable lesson, Sheppard says. "If you introduce something without any regulation and give people the legal freedom to do that, then you're going to set yourself up for huge long-term consequences and costs." Now, he notes, most countries have a list of species that are not allowed inside their borders. Australia has a list of prohibited species, but it also has an additional list, a short list of live species that are allowed. All others need not apply.

Cane toads seem like a good example of scientific hubris— introduce something to control a pest, and it becomes a pest itself. But their story also shows that, in the long run, the ecosystem was far more flexible than assumed. And they show that when we try to sledgehammer our way through ecosystem problems we don't understand—by adding species or by trying to kill them off wholesale—the hammer might well land on our toes.

WHEN CHINA ENGINEERED a mass effort to eradicate the Eurasian tree sparrow, the consequences were dire. But they did actually get rid of the sparrows, something that French North Africa could only have dreamed of doing in the nineteenth and twentieth centuries. French colonists in Africa had ended up with a plague of sparrows—one they basically brought on themselves.

The culprit in this case was the Spanish sparrow (*Passer hispaniolensis*). Unlike Eurasian tree sparrows or domestic house sparrows (*Passer domesticus*), some Spanish sparrows are migratory. Every year, they fly down to North Africa, to places like Algeria, Tunisia, and Morocco, where they nest, rear their young, and then head back to Spain and Portugal, having eaten the area pretty much bare in the process. A British scholar visiting Morocco in the 1890s wrote that the area was "almost devoid of cultivation and covered with dense undergrowth of arbutus and other bush. The reason that this land is uncultivated is that the place swarms with wild boar and sparrows, one of which uproots the grain, while the second destroy the ripening ear and leave not a particle." The French, apparently, didn't consider the sparrows important when they began to colonize North Africa in the nineteenth century. But, boy, did they soon find out.

"They, like many other people in Europe at the time, had been reading all these Roman-era texts that talked about the fertility of North Africa," explains Jackson Perry, a historian at the New York Botanical Garden. North Africa had been the breadbasket of the Roman Empire, feeding the legions as they tramped from Britain to the Middle East and back. Almost two millennia later, France sought to follow the Roman lead.

When the French colonized Algeria, they dismissed the small farms and sheep and goat herds of the North Africans, blaming them for not taking advantage of the soil's capabilities, Perry says. Instead, the French planted monocultures of barley, wheat, and oats, and separated the fields and marked their property lines with lovely windbreaks of imported eucalyptus trees.

The eucalyptus trees weren't just an aesthetic choice. "They thought it could prevent malaria," Perry explains. Eucalyptus trees do not prevent malaria. But if you believed in miasma theory—the idea that icky smells carry disease—eucalyptus seems reasonable. They smell nice—no miasma here. And the trees are very thirsty, sucking up standing water. "If you plant a whole lot of eucalyptus trees in a marshy area, it will reduce the amount of mosquito breeding grounds that there are."

But the pretty, healthful trees turned out to be a fatal mistake. The Spanish sparrows adored them—nice high spots for nests that no predator could reach, and planted right next to huge fields full of grain. It must have been like getting an apartment over a bakery.

Perry was actually studying the history and spread of the eucalyptus tree, but in the process, ended up reading the many complaints about sparrows from the French colonists of eastern Algeria. In Chlef (then called Orléansville), the military governor complained that sparrows ate up an entire eighty-hectare barley plantation next to a tree nursery. The governor estimated that the nursery trees had 284,000 sparrow nests.

French law protected wild birds, but locals soon got permission to take matters into their own hands. They started hunting down sparrow nests—often hiring young boys to do it, who could climb higher up in the trees. "The colonists wage war incessantly; they feed on sparrow dumplings, and sparrows' eggs, they throw sparrows to their dogs and pigs, and yet 'the cry is still they come,'" a forestry journal reported. "The French have conquered the Arabs and won their colony, only in their turn to be conquered by the sparrows." The town of Guelma reported killing three hundred thousand sparrows a year, only to see more come back the next.

Labeling something a pest means you can do almost anything you like to it. M. A. Tazi, who would become a famous Moroccan filmmaker, wrote about his childhood sparrow exploits. "We would climb up the eucalyptus trees and we'd gather the little infant sparrows—they were no more than little balls of feathers and couldn't move or fly or anything," he wrote. "We'd gather them and lay them out in a line on the road, waiting for a car to come. We called this 'the machine gun game,' after the sound made as the car crushed them. For us, these sparrows had to be destroyed and it was best to destroy them at the very beginning, while they were still in the nest when we could catch them. Now, when I think of it, it's a horrendous sort of cruelty."

The French knew full well what was happening. They realized very quickly that the sparrows were using the eucalyptus trees as a home base to eat all their crops. You'd think that the colonists might just cut down all the trees. But it wasn't that simple. The trees were on private land and grown by private people or companies. "These are communal problems," Perry says. "And so it will be difficult to say to a single landowner . . . or to the railroad, 'okay, you have to cut your trees down.'"

Algeria gained its independence from the French in 1962. But they still aren't free of the sparrows, which continue to munch on the grain fields. Or of the eucalyptus trees, where the sparrows still nest. The government continues to organize antisparrow campaigns to this day.

THE FRENCH, ALGERIANS, and Chinese are not alone in wanting to be rid of sparrows. Several colonized countries have waged sparrow wars—often after originally introducing the birds on purpose.

European acclimatization societies, like those that brought the rabbit to Australia, brought European sparrows from Europe to North America and Australia for a little sense of home. In the nineteenth century, sparrows and other animals from Europe were a way to impose will—ideas of what the right birds and mammals looked like—onto new, strange places. If it was an exercise of power to try to get rid of the sparrows, it was an exercise of power to introduce them as well. The small comforts of home—even the cheep of a sparrow—gave the colonizers a little of their confidence back.

But at what cost? The clever, adaptable birds spread wherever they landed. Soon scientists were waging wars in journals—and other people waged war in newspapers—over whether a sparrow was friend or foe.

The countrywide Chinese mobilization to get rid of the birds might seem extreme, but it did have a basis in a pretty reasonable fear, explains Fa-Ti Fan. During the Korean War, northeastern China suffered a plague of dead rodents. Many people, even U.S. sympathizers, thought it wasn't impossible that the United States might have deployed some form of biological warfare, trying to spread a disease into the Chinese population via its rodents. (The United States denied it. An international investigation concluded the opposite in 1952. Subsequent document releases from the then Soviet Union showed the accusations were made up. But there is still controversy.) This gave Mao a powerful license to get rid of pests, or anything that might carry a disease.

The sparrows were more of an agricultural pest than a public health menace, but they did have some scientists backing up their status, Fan says. One scientist, the father of Chinese ornithology Zuoxin Zheng (sometimes called Tso-Hsin Cheng), was assigned to study their impact. He and his team fed some caged sparrows, to see what they ate and how much. The scientists decided the sparrows ate mostly grain, did some math, and concluded they could be harmful to the country.

Zheng originally made the claim that sparrows destroy 35 percent of rice ears, the historian Weimin Xiong notes. Despite this, Xiong writes, Zheng did not think that all sparrows needed to die. "He believed that only a few sparrows should be killed at certain times." But the Communist Party's Central Committee was not enthusiastic about moderate measures, and Zheng did not want to go against them.

Other scientists weren't satisfied with such back-of-the-birdcage calculations, arguing that birds ate both insects and grain, and could save crops from bugs as much as steal the crops themselves. "They are going back and forth debating about this issue," Fan says. "On the other hand, though, Communist political policy is becoming more and more clear, that they want to put this bird into the category of bad, harmful pests, right? And then it becomes more difficult for people to speak up."

As people began to report insect infestations in 1959, however, the scientists spoke up again, Xiong writes. Not only did the scientists argue that sparrows had been unfairly demonized, they said the answer to the insect problem was more sparrows, not less. "A single bird is more effective at catching insects than one person or even one hundred people," Xiangtong Zhang, who was a neuroscientist and member of the Chinese Academy of Sciences, was translated as saying. "The biological approach to killing insects, namely using birds, is an excellent option."

Zheng agreed, noting that he'd never wanted the sparrow gone entirely, only sparrows that were pests. He also connected recent insect infestations to the sparrow campaigns, and encouraged the Party to change its mind, Xiong wrote. Sparrows were finally exonerated. Zheng, however, ended up punished by the Party during the Cultural Revolution for his intellectual work.

The sparrow blitzkrieg wasn't about science at all. It was about Communist China—who it wanted to be, and what it wanted to achieve. The pest campaigns weren't about pests, but about empowering people, Fan explains. "Controlling nature also means that we control our destiny." The Party wanted to give a feeling of power to the people, and it did that, in part, by giving them campaigns to control their environments.

Killing all the sparrows, rats, and mosquitoes in China was just one example of "the extreme degree to which we see all kinds of things being taken amid the Great Leap Forward," explains Mary Brazelton, a historian of science at University of Cambridge. Lots of things during this era in China—before and during the Great Leap Forward—got turned into mass campaigns. Not all of them turned out badly. Vaccination campaigns in the early 1950s aimed to get everyone vaccinated against diseases like smallpox—and succeeded. With early success against diseases, it's easy to want to expand to killing sparrows, smushing mosquitoes, and even collecting scrap iron en masse. The emphasis, Brazelton says, was about success, accomplishing everything for the sake of showing that China was going to realize Communism faster and more decisively than everyone else. "It's the same idea of totality, in a sense of achieving total coverage of an area in terms of vaccination," she says. "It's construed in an interestingly positive sense."

With so many campaigns trying to get to 100 percent, it was probably not the eradication of sparrows alone that resulted in the deaths of anywhere between fifteen and fifty-five million people, says Miriam Gross. There were plenty of other contributors. Grain production demands were increased, and with them, a lot of Communist Party peer pressure. "There is an incredible pressure, peer community pressure and government pressure and fear of the last campaign, to make sure that your area sounds like it's producing at least as much as everyone else," Gross explains. The local leaders didn't want to say they produced too little grain—but also didn't want to say they produced too much to keep expectations low.

The net result is that everyone slightly overinflated their numbers. Those got passed up the chain, where they got inflated again, and again, and again. At the highest level, the government then set the taxes for each area—taxes calculated based on the reported grain numbers. Farmers ended up getting taxed—and paying for it, in grain—on the basis of grain they never produced in the first place. Some Party members then tried to hide famine evidence from their superiors so that they would not be punished.

"I mean, it's the agricultural era, they are producing this tiny surplus," Gross says. "And if you tax away that amount, you're dead in the water." As proof of Communist success, China also exported large amounts of grain to other countries, to prove just how much food they (supposedly) had. The government also required more from its people, starting huge building projects of roads, bridges, and dams. That required manpower—men who otherwise would have been in the fields getting the harvest in. Villages were told to bring in and smelt down all their scrap iron in backyard furnaces—often sacrificing farming implements. At the beginning of the Great Leap Forward, communal dining halls promoted large amounts of food waste, decreasing food supplies. Fields were allowed to lie fallow. There was drought in some places, and huge floods to others.

Gross and other historians don't think the sparrows were the overriding cause behind the great famine. Politics did that. But the lack of sparrows didn't help. When people are living from hand to mouth, a few million sparrows eating insects can make the difference between life and starvation. In the Four Pests campaign, an effort to get rid of a pest that was supposedly causing hunger left famine in its wake.

Without persecution, the sparrows returned, Sun says. They are common around his hometown to this day.

CHINA AND FRENCH-OCCUPIED Algeria are far from the only countries to try to get rid of sparrows. Under Frederick the Great of Prussia, there was a renewed effort to slaughter sparrows in 1744

(Prussia had already tried it starting in 1656). By the 1790s, Prussians began to realize that without sparrows, caterpillars would eat their fruit trees bare. Australian colonists, having introduced sparrows, then tried to kill them off, using poison, guns, and cats. The cats then went on to become their own problem, as we have seen.

In so many environments, humans try desperately to introduce, remove, and control the numbers of one species or another. We are flexing our muscles, using our power to alter the landscape for our own ends.

When people—often people in a land where they've taken power—use that power in an ecosystem they don't fully understand, they can end up with insect invasions, cane toad catastrophes, and swarms of sparrows. And often it is because the people involved, the colonists, the political parties, didn't listen. They didn't listen to scientists, and sometimes scientists themselves succumbed to pressure. They ignored local people and their knowledge, convinced that they knew best. Blinded by what they wanted and believed, they were forced to learn the hard way, over and over, confronted by pests of their own design.

PART V

THE ONCE
AND FUTURE
PEST

9

A HERD
OF DEER

My Aunt Carrie used to think deer were cute.

She and my uncle have lived in the suburbs of Atlanta for the past thirty years or so, where they've raised my two cousins and a small fleet of miniature dachshunds. The neighborhood is what passes for a leafy one in Georgia—which means it's mostly pine with a healthy mix of oaks, maples, and other hardwoods. But there are also creeks, landscaping, and plenty of grass. It's a lovely spot for deer.

Every morning, before my aunt lets the dogs out into her fenced backyard, she walks the perimeter, checking for deer. Her wiener dogs have what can only be described as a crazed reaction to hooved herbivores. They see a deer when walking around the neighborhood and morph into tiny, raging, snarling hunters, intent on running down something one hundred times their size on their stubby little legs, and then, presumably, barking it to death.

To avoid deer dying of burst eardrums, Aunt Carrie checked her three-foot-high fence. Until one day in spring 2020, when she didn't. Her then twelve-year-old dog, Bowie, bounded joyfully out into the yard and made a beeline for the back of the nearly half-acre property.

"He started barking in a different bark," my aunt told me. Carrie headed over to see what he was doing. Their half acre is shaped kind of like a pizza slice, with the house at the crust. The front half of the backyard is open, with a fire pit, table, and landscaping with paths. The back half is wooded, descending down to a stream and tapering to a point. It was toward the back half that Bowie had gone.

Carrie found her dog barking madly up at a large white-tailed deer (*Odocoileus virginianus*). As she watched, the doe reared up and crashed down. Bowie yelped and fell. My aunt wasn't sure if the deer had actually stomped on her dog. The doe ran off, and my aunt raced the dog inside.

Bowie appeared to have all four tiny legs intact and lay down easily enough. Later, however, she saw he'd developed one gigantic bruise down his entire doggie belly. Whether the bruise was from deer hoof or the fall, he'd clearly had some damage. His reactions to deer became, if possible, even more extra, and Carrie got even more zealous about checking the fence.

But in spring of 2021, around 7:30 in the morning, her newest hot-dog-shaped family member, two-year-old Deegan, slid through her legs before she could do her daily perimeter pace. He darted into the backyard and started the same crazed barking. "It's a very different bark than him just barking up at a squirrel," Carrie says.

My aunt raced to the scene to find Deegan furiously trying to get at the legs and tails of three fawns, tucked in a row under her pampas grass. A doe circled nervously nearby. Carrie saw the dog's life flash before her eyes. "I'm screaming and crying because frankly, I'm seeing that my dog is going to be dead," she says. She frantically attempted to grab Deegan while trying to keep away from the mother deer. Deegan thought this was all a marvelous game and refused to be caught. My aunt, screaming madly at both Deegan and doe, eventually managed to snag the dog.

Aunt Carrie had started noticing the deer a few years ago, near the tennis courts where she and my uncle play doubles with other people in the neighborhood. At first, "you might see three or four" when they played in the mornings, Carrie says. "Now we see twelve to twenty."

The landscaping and tasty neighborhood gardens might be enough to explain the increase, but that's not all. One of Carrie's close friends feeds them. Every morning, she puts out large metal vats full of sweet corn that she buys from a feed store. Several other neighbors delight in feeding the deer and welcoming nature into their yards. The net result, of course, is that the deer also end up as nature in other people's yards—nature they do not always appreciate.

Carrie had made previous concessions to the deer. She no longer planted any flowers or hostas, knowing deer liked to eat them. Two raised garden beds were replaced by the fire pit. She could never grow anything because the deer ate it all.

But after the second dog-deer showdown, she'd had enough. My aunt isn't just on the homeowners' association board; she's also the president. At her request, some of her friends wrote to the board with deer complaints, and Carrie asked that the board consider a ban on feeding the deer.

Her deer-feeding friend is also on the board and stated she would not stop putting out vats of deer feed anytime soon. The neighborhood did what middle-class suburbanites do best—they avoided the issue completely. "The board felt that it wasn't really anything they wanted to tackle." Carrie, unwilling to risk a rupture with her friend over an herbivore, went to plan B. She spent seven thousand dollars on a six-foot-high fence—the highest the board would approve.

I hated to tell her, but six feet isn't high enough. A typical white-tailed deer can leap eight feet. She still checks for deer every morning. "I used to think they were so freakin' cute," she says. Now, she just sees them as a threat.

Most people who encounter deer probably don't end up treating bruised doggie bellies. But they do end up seeing them in their yards and parks, or as bloated dead bodies on the roadside—or possibly in front of their own headlights.

Between 2019 and 2020, 1.3 million people in the United States billed State Farm auto insurance for crashing into a deer. Americans have a 0.6 percent chance of hitting our hooved neighbors on a winding country road or a busier four-lane highway every year—or at least, of filing an insurance claim for it. In fact, the most dangerous animal encounter in North America by the numbers isn't a wolf, bear, or great white shark. It's plowing into a white-tailed deer.

Those car crashes have consequences. Thousands of people in the United States are injured every year when their cars collide with an ungulate—a large hooved mammal that could be a deer, elk, or (God forbid) a moose. Some researchers estimate 440 people die every year. That's almost a billion dollars in damages, thousands of hours of lost work, and untold physical and emotional pain.

That's before counting the crop damage. Deer love what we plant, whether it's our tasty dogwood and yew trees or the rows of corn, alfalfa, and soy that feed us and our farm animals. Dollar amounts of just how much deer eat per year in the United States are usually taken locally and are difficult to calculate, but the National Agricultural Statistics Service estimated in 2001 that around $765 million dollars' worth of field crops, nuts, berries, and veggies vanished into the hungry stomachs of wild animals. The service estimates that about 58 percent of the damage to field crops and 33 percent of the damage to nuts, veggies, and fruit was probably due to deer. And that's not even counting the deer looking guilelessly up at you from your home garden, mouth stuffed with your baby kale.

Most people probably don't see a deer, or two, or ten and think, "Ugh, this is all our fault." But it is. From near extirpation at the end of the nineteenth century, deer have bounded back into our lives. They've done it because we did more than passively let them. We actively encouraged their return. We regrew young, green forests, set aside plots of land where deer would never be hunted, built subdivisions full of tasty plants, and landscaped with deer candy.

We created the ultimate deer habitat. In return, the deer proliferated. Now, in many places, they are eating themselves out of house and home, transforming our forests into something previous generations of deer and humans alike would not ever have recognized. As the East Coast is faced with habitat destruction, car crashes, and sick and starving deer, its residents are faced with having to control—or even kill—the consequences of their actions.

BEFORE THE ARRIVAL of Europeans, an estimated thirty million white-tailed deer roamed North America, but even on the East Coast, they didn't roam equally everywhere. "There is a bit of heterogeneity in what that landscape actually looks like, on the eve of European colonization," says Elic Weitzel, an archaeologist and human ecologist at the University of Connecticut, who has been studying evidence of deer populations under Indigenous management. "Indigenous folks are burning and managing the landscape," he says. But "they're not going to do that everywhere. They're going to do it close to where they live."

Native Americans across many cultures hunt and eat deer, and burning was a common management strategy to encourage grass growth. Where grass grew, deer followed. While many people think of deer as creatures of deep forest, that's not quite true. Deer prefer to eat seedlings, buds, small branches, fruit, and forbs—herbaceous plants with tasty leaves. But there are relatively few succulent things to eat in the truly deep woods. Acorns in mature forests are good eating, but really old, mature forest has large trees that block out most of the light and relatively few seedlings or tasty cover. With burning, Indigenous people opened up the ground to the sky, facilitating lots of new plant growth—lots of food for deer to eat.

Among the Seminole in Florida in the late nineteenth and early twentieth centuries, for example, deer was one of the most common food staples. This was partially due to preference, but once leather-loving Europeans arrived, it was also due to the value that hides had in trade.

With the Europeans, things took a dark deer turn. Colonists liked venison, deer hide, and deer hair, too, and slaughtered the animals across the East Coast in the seventeenth and eighteenth centuries. It got so bad that when cold winters also conspired against the deer in 1698, Connecticut had to reduce the length of the hunting season. Additionally, in the seventeenth and eighteenth centuries, the vast majority of the northeastern United States was logged for firewood, houses, fences, and barns—and to clear up land for agriculture. Habitat and hunting took their toll, and deer suffered a huge drop in population numbers.

Along with killing off the deer for our venison dinners, our efforts killed off wolves and mountain lions, two of the main predators of deer. Soon, people were the only major deer predators left, and our appetites were voracious. By 1850, Kurt VerCauteren, a researcher at the United States Department of Agriculture, estimated the population had dropped below fifteen million. In 1900, there were not even a million white-tailed deer left in the United States.

By then, colonists had somewhat belatedly realized their mistake. Rich people began buying land and setting it aside to become "wild" (for their own hunting pleasure, of course). They began importing more white-tailed deer from remnant populations like those in the Adirondacks. States began to strongly regulate hunting.

In 1900, a law known as the Lacey Act made it a federal crime to poach and sell wildlife across state lines. That made a lot of "market hunting"—hunting to sell meat, fur, and feathers—functionally illegal. And as the Industrial Revolution and then the Great Depression hit the United States, people left the clear-cut former forests and moved back to the cities. Woodlands began to grow up where fields had once covered the landscape.

Deer populations were also actively encouraged by agencies, in part to allow for more sport hunting. "I call it America's deer industrial agency complex," says Thomas Rawinski, a botanist at the Harvard Forest, "because the state agencies that manage deer on our behalf, supposedly, they also profit from lots of deer, they sell more licenses, more permits." There's plenty of businesses involved, too, selling apparel and guns and bows and gadgets, but they don't directly keep the deer on the landscape. The state agencies do. "The agencies . . . are facing declining hunting numbers, and aging hunters, and they don't want to do anything that might threaten the last remaining hunters, and they're trying to promote hunting," he says. "And one way to do that is to help ensure that there's plenty of deer to shoot."

In the twenty-first century, deer live where we encounter them: in secondary growth forests. And wow, we have grown so many of those.

People often think they need to drive to find woods, but for about two thirds of the United States population, they actually live in the woods already, explains Jim Sterba in *Nature Wars: The Incredible Story of How Wildlife Comebacks Turned Backyards into Battlegrounds*. "If you draw a line around the largest forested region in the contiguous United States—the one that stretches from the Atlantic Ocean to the Great Plains—you will have drawn a line around nearly two thirds of America's forests (excluding Alaska's) and two thirds of the U.S. population," he notes. Sure, it's not the endless march of green and brown we are told to picture when we see a forest. Instead, it's pockmarked with buildings and latticed with roads. But it is forest nonetheless.

This forest—the trees in our neighborhoods, and the woods we hike in, and the nature we adore in places like Shenandoah National Park, the Great Smoky Mountains, and the Adirondacks—is mostly secondary growth, filled with relatively young trees. These leave nice big gaps in the forest canopy, allowing in beautiful shafts of sunlight. The sun in turn allows the growth of seedlings and forbs—deer food.

Since about the 1970s, deer populations have increased. Now, the nearest estimate is that more than thirty million deer are picking their way delicately through the undergrowth around U.S. homes and businesses, slotting themselves into the habitat Americans made. Those deer are not equally spread out, though. By 2008, some areas, such as most of Maine and southern Florida, had fewer than fifteen deer per square mile. Others, like northern North Carolina and most of Wisconsin, had more than forty-five deer per square mile.

This is great, in theory. Hunters wanted a lot of deer, so deer were preserved. More deer is exactly what they got.

NO ANIMAL FLITS through a landscape without changing something. Humans, of course, change a lot. We trample paths, put down concrete slabs, plant crops, lay asphalt, and alter our landscapes in a million other ways. But there's more than one way to be an ecosystem engineer. Beavers build dams, creating ponds and wetlands that drive people who built a road on land they thought was dry pretty nuts. And ungulates vacuum up plants, leave poop, and run.

Horses (*Equus caballus*) love to run in herds. When they do, they throw the weight of hundreds or thousands of ungulates onto the ground—over and over again. In some areas like the Great

Plains, the prairie ecosystem has formed itself around the movements of big hooved grazers—bison (*Bison bison*). Bison trample down and pack prairie soil, and then poop on it, restarting growth from the ground up—much in the same way fire does. The young, tender shoots that take over make more food for bison and other animals that live on the prairie. The grasses that create prairie sod even may have coevolved with large grazers. The flat, trampled areas also provide new opportunities, such as places for turtles to sun on lakeshores.

But not all soils are meant to be sod. Especially not some of the soils in Australia.

Horses arrived in Australia with the first European colonists in 1788, and were escaping (or being abandoned) and running feral by 1804. Australia—full of wide, flat places to run and grasses to eat—suited them well. Now, around four hundred thousand feral horses roam Australia. Most are in the Northern Territory, but a herd of around twenty-five thousand roams the Australian Alps in New South Wales and Victoria.

White-tailed deer do have cultural value (here is your obligatory *Bambi* reference)—but they've got nothing on horses. Every good knight has a mighty steed. Every wild and free young child has a Black Beauty or a Misty of Chincoteague. Every midlife crisis has a Mustang. In Australia, they call feral horses "brumbies," and the animals are especially famous. There's an 1890 poem by Banjo Paterson that became a movie in 1982, called "The Man from Snowy River," in which an expert rider rounds up a herd of feral horses. The poem is wild, brutal, and beautiful, and reads like it could be chanted or sung:

Then they lost him for a moment, where two mountain
    gullies met
In the ranges, but a final glimpse reveals

On a dim and distant hillside the wild horses racing yet,

With the man from Snowy River at their heels.

There's also a famous series of children's books called *The Silver Brumby*, about a silver horse named Thowra and his adventures in the Snowy Mountains.

Now, many of those Snowy Mountains are Kosciuszko National Park in New South Wales. It's named for its tallest mountain, Kosciuszko, which features in "The Man from Snowy River." The brumbies there are more than feral horses. They're a tourist draw and symbols of starry-eyed freedom to horse-loving kids and adults.

Maggie Watson, the conservation biologist at Charles Sturt University previously seen studying pigeon poo, wants to have the brumbies shot at from helicopters. This is called aerial culling, and lest you react with shock to this proposition, it's a fairly common way for people to deal with large populations of skittish animals. The United States is in on it too. You can, in fact, book a team-building exercise with your business to shoot feral hogs from helicopters in Texas. God Bless America.

Aerial culling already takes place in northern Australia for horses, buffalo, and wild pigs, and in central Australia for camels. In New South Wales, however, it's banned. In a previous aerial culling attempt in Guy Fawkes River National Park in 2000, the Royal Society for the Prevention of Cruelty to Animals launched a legal campaign against the National Parks and Wildlife Service for leaving badly injured horses in their wake (an investigation later found one horse had survived for two weeks with gunshot wounds). Since then, New South Wales has banned the practice. Watson and other conservationists want the copters brought back.

When Watson pops up on Zoom, she turns her camera toward the paddock near her house in Albury-Wodonga, where her lanky pinto, Lola, is being introduced for the first time to her new friend, a smaller roan horse named Lotta. Things are so far going well; there are a few tiny kick attempts, but neither horse really means it.

What Watson wants to show me, however, is the paddock itself. She turns around and focuses the camera on a path through the paddock toward the water trough. "They do this just like sheep, you know," she says. Horses take the same routes over and over again, trampling paths in the landscape. "It makes sense from an evolutionary point of view, because they've got these long, delicate legs. . . . So they make these paths for them and for their young so that if the big bad mountain lion comes, they can run straight forward and know that their hooves aren't going to fall into a hole and they're going to die."

While the soil in Watson's paddock is one thing, the soil up in the Australian Alps is another. Kosciuszko National Park, as well as other nearby parks, contains alpine sphagnum bogs. These bogs are "endangered ecological communities" in Australia and contain a large number of specialized plants. They also play host to highly specialized animals such as the southern Corroboree frog, a tiny yellow-and-black poisonous frog that lays eggs in tiny cups of moss, eats ants, and is found nowhere else in the world. There are also specialized fish in the fens and streams, lizards, and freshwater crayfish.

What the bogs (and the Corroboree frogs) aren't used to playing host to are horses. "It wouldn't be an issue if this continent had evolved with animals with hooves," Watson says. Then, the plants would have evolved to deal with the packed soil and intense nibbling that comes with ungulate herds. But Australia had only the (relatively) soft padded feet of kangaroos and wallabies. The plants never saw the brumbies coming.

The feral horses galloped into the Australian Alps—and over the tiny frogs. "At this point there are—and I kid you not—fifty of these frogs in the high country. Fifty. Five Zero," she tells me.

The brumbies also poop in the waters of the fens and streams, endangering fish. They eat up sensitive and endangered plants.

Stuck in an alpine ecosystem as they are, the sphagnum moss denizens have nowhere to go. "You just can't have this destructive animal and the natural wildlife that evolved there, simply because there's just not enough room," Watson says.

It might seem a bit odd to have feral horses—horses that are definitely not native to Australia—protected in a national park. It was so odd that Maggie Watson's husband, David Watson, a professor of ecology at Charles Sturt University, resigned from the New South Wales Threatened Species Scientific Committee in protest. "We'd just gone through final edits for listing feral horses as a key threatening process," he says, when the federal government began putting through the 2018 Kosciuszko Wild Horse Heritage Bill that protected the horses in the park. David Watson felt as though his opinion and the weight of the scientific evidence was being tossed aside. Since then, Watson and Watson have been speaking out against the presence of feral horses in Kosciuszko National Park as the herd numbers have continued to increase.

Maggie Watson is especially hot under the collar about Arian Wallach, an ecologist at the University of Technology in Sydney, who is involved in the movement of compassionate conservation— the idea that conserving populations and species alone isn't enough, you have to consider the welfare of the individuals in the populations as well. Wallach is Team Brumby, and against the move to have them culled.

Wallach does not dispute that brumbies cause change. But she doesn't think ecologists can say that the change is a bad one. "I deny that science can measure or define that negative ecological impact," she says. In other words, it can measure ecological changes or effects, but it cannot measure whether or not those changes are bad.

In addition, the frogs aren't just suffering from an overabundance of horses. The Watsons acknowledge that the frogs and the bog areas are suffering under a variety of pressures—climate change being a major one, and the chytrid fungus that has been infecting amphibians all over the globe another. In fact, wildfires burning through in 2019 and 2020 nearly doomed the frog entirely. Wallach is not convinced the horses are the biggest danger.

Some people would say that these horses are here because we brought them here. It's therefore our responsibility to get rid of them. Wallach isn't so sure. "It's this idea that you can regard something like a horse as you would plastic pollution, you know, we put the pollution in the ocean, therefore we're going to clean it up," she says. "And that's the responsibility. There's no one there. There's just pollution." But, she says, horses aren't oil or plastic. They are living things. Living things who may have escaped human hands, but who came to the Australian Alps on their own.

Many of the native residents of the Australian Alps would probably be better off if the horses were not present. But some fans of the brumbies would rather they not be killed, and instead be moved to another location, to live out their lives as domesticated horses.

Maggie Watson points out that moving large numbers of horses is no easy matter. "As soon as you get horses together in a group, they start kicking," she says. "And you end up with a lot of animals with broken legs that you've got to shoot." The rest might well be too full of worms or with long-term injury, making them very difficult to adopt, meaning most of the brumbies end up relocated to a slaughterhouse in South Australia, Watson says. It is simpler and cheaper to cull in the first place.

Wallach isn't swayed by those arguments. "Let's say I was convinced for the sake of argument that horses were a major cause of a particular frog species going extinct," she says. "Well, then, you know, geez! We're creative, innovative animals, we'll figure something out!" We can be better, she insists. We don't always have to go to war with nature that displeases us—no matter how scientific that displeasure may be.

Wallach's ideas are controversial to some Australian ecologists.

According to the ideals of compassionate conservation, Wallach disagrees with conservation programs to harm wildlife—from the carpeting of remote islands to kill rats that threaten seabirds to the shooting of brumbies to protect the Australian Alps.

This is at odds with many conservationists such as Maggie Watson, who place more value on the biodiversity of species than on the well-being of individuals. In a 2019 paper, Watson argued that letting invasive species like brumbies coexist in alpine bogs would lead to the destruction of horses and frogs alike. A failure for compassion and conservation. To Watson and other conservationists, killing horses or rats is painful, but for the sake of saving species, it is necessary. It's the conservation mindset that governs programs like Aotearoa/New Zealand's Predator Free 2050, and the programs to control rabbits, cane toads, and now horses in Australia.

As Wallach speaks with me, she pauses suddenly. She's spotted a tiny frog—a dainty green tree frog (*Ranoidea gracilenta*). She holds it on her hand, marveling at its beauty.

One ecologist wants the horses shot, and shows me her love for horses. The other says maybe it's natural for frogs to go extinct, and pauses to admire a frog. They are both hammering me over the head with their messages—it's not the horses or the frogs. It's the ecosystem. One wants to save the ecosystem they are losing, because in conservation, biodiversity is the most important thing. The rare above the common, the native above the invaders. The other wants to know why we need to kill other animals to do it, animals that can suffer individually just as much as the native animals can. Why we can't accept that change happens, and ecosystems won't stay the same.

The current compromise pleases no one—the horses and frogs included. The last fifty frogs are preserved behind a fence. Conservationists released more frogs in 2022 to increase population numbers. It's not enough genetic diversity to sustain the population, David Watson admits. Other species are probably doomed with them. Meanwhile, the latest plan is for culling the horses by a few thousand with people on foot.

It's important to note here that this isn't the whole population of wild horses in Australia. The Watsons and other conservationists object to the ones in Kosciuszko National Park because they are in a national park—a place that is supposed to be dedicated to the preservation of biodiversity.

The trick will be managing the feral horses in New South Wales so they stay off the sphagnum bogs—and still look free enough to give people their wild brumby experience. "How do we manage welfare issues with these animals?" David Watson asks. The horses get sick. They step in holes and get hurt. If they're domesticated horses, they should probably get veterinary care. But if they're wild, shouldn't they be left to suffer and die in peace? "How managed can they be," he asks, "and still have a wild experience for people?"

WHERE THERE ARE too many deer, many conservationists believe that looser hunting seasons and limits, or large, controlled culls like those for horses, are the way out of the problem.

Nonhunters might think hunters have it easy. Deer in suburban environments are used to people to the point of hubris. I've run past does standing so close to trails that I've almost wanted to reach out and spank them—just to give them a tiny bit of fear for the human stream going by. For those used to the presence of deer in suburbs, bedding down openly in their backyards, it seems all a deer hunter would have to do is aim a rifle out the back door to hit something.

Don't be too sure. I'm sitting on the ground, back to a tree in Fauquier County, Virginia, wearing blaze orange. (Fun fact, deer do see color. They can even see orange, though they don't see it as the vibrant hue that we do. Luckily, other rifle- or bow-toting humans see blaze orange as blazing in everything but darkness.)

It's around 40 degrees Fahrenheit, I've been up since 4:30 in the morning, and I'm so bundled up I can barely move.

I'm next to my friend Crystal Lantz, hunting on her relatives' property. Crystal and I went to high school and then college together. We even got PhDs in related fields. In all that time, I only vaguely knew that she spent some weekends out hunting. She shoots skeet when deer and turkey are out of season.

Crystal has been hunting all her life. "I've been handling guns since I was like very small," she says. Hunting has helped to fill her freezer and feel close to her family. "I do really enjoy venison," she says. "But it's also something about staying connected with all of my family's roots back there. So many people for so long were subsistence hunters. If you want to eat meat, that's what you did." Now she's a silent presence seated at my right, a hand placed carefully alongside the trigger of her Weatherby 6.5 Creedmoor, with the tip of the long barrel perched on the toe of her boot.

Together, we quietly watch darkness give way to dawn. At first, it feels like your eyes are adjusting to the monochrome dim, the pale of old stone boundary fences, the dark pillars of trees holding up a charcoal gray sky. Then you blink, and suddenly the holly bush to the left really is kind of green. You blink again a few minutes later, and the blackberry canes are reddish now. The dirt is brown, the leaves different tones of tan and rust. Blink again. It's morning, and even colder than before.

I suddenly get why people do this. It's perfect. When you hike, you're surrounded in a bubble of your own sounds, your breathing, your clothing, and your steps. Here, it's so silent you can hear a bird fly, the *whapwhapwhap* of its wings thrusting through the air above your head. We sit so still a small animal scurries in a pile of leaves to my left, its tiny back making the leaves vibrate four inches from my foot. I feel absolutely present in the moment, the way I can never feel when meditation instructors coo their smooth tones into my headphones.

The distant roar of the planes from Dulles and the rumble of nearby highways suddenly sound like the most selfish things in the world. How dare we make that kind of noise, when morning can bring you this?

We are not in the suburbs. We're further out than that. Human sound just travels very, very far. Hunting deer in the suburbs, Crystal explains, violates a lot of hunters' feelings of "fair chase." It's not fair to hunt deer so incredibly habituated to humans. Fair chase, she says, means you have to earn it.

For something called "fair chase," there's a lot of sitting still. And while I love this, and I want an excuse to come out here in the dawn every day, I also understand why hunters want to see more and more deer. We've seen plenty of squirrels, some birds, and a dude checking his camera traps. We spot something that could maybe be a coyote. My butt has been numb for two hours. Crystal raises her rifle once. Nothing. Most mornings hunting is like this. A few hours freezing, and then back home with your now tepid coffee. For a landscape in which we are the only predators left, we're doing a terrible job.

But maybe expecting a deer to just saunter into our sights is the wrong way to think about it. "From a tribal perspective, you don't really take an animal's life, so much as you accept it giving its life to you to meet your need," explains Peter David, a wildlife biologist with the Great Lakes Indian Fish and Wildlife Commission. He is not Indigenous but is hired to assist the members of the Ojibwe tribes with their off-reservation territory rights. "It's not a given that there has to be a deer behind every tree." When nontribal members don't see deer, he says, they tend to think there's something wrong with the deer—especially if they are baiting and expect the deer to come to them. But if tribal members don't see deer, they assume that maybe they just didn't go to where the deer were. It's a failure of the hunter. Not of the deer.

The tribal perspective also doesn't see other predators, such as wolves, as competition, he says, but as important partners in keeping the balance of deer on the landscape. People "kind of pooh-pooh the tribal perspective sometimes as being just a cultural thing. It overlooks the fact that the tribes lived here for centuries and centuries and centuries." The Ojibwe, he notes, have very practical knowledge of how deer impact the ecology around them. "There's plants that are gathered for food or for medicines that can be negatively impacted by an overabundance of deer. And so one of the basic terms about the good life and the tribal perspective is having things in the right balance. And I think [tribal] people are willing to see that balance rather than sort of like, deer above everything else."

Balance means wolves and humans, both hunting to keep the deer down and the plants alive. Forests managed by the Ojibwe and the Menominee have tolerance for predators. They also have more plant diversity, better tree growth, and lower deer densities.

Without enough hunters, deer proliferate on the landscape. This is bad for us. We hit them with our cars, suffer stripped gardens, and sometimes they bruise our dogs. But it's no good for the deer and the environments they live in either.

Unlike wild horses in Australia, deer are native to the United States, but not in the concentrations that there are now. There might be just as many deer as there were before Europeans came colonizing, but those deer are packed into smaller areas, hemmed in by roads, concrete, and many, many people. Those deer have to eat. According to Tom Rawinski, the botanist at the Harvard Forest, the average deer needs about seven pounds of forage per day, and they're not getting it from nice calorie-dense stuff like cheese. "I think a researcher once intentionally clipped little twigs, to see how many twigs it took to comprise ten pounds of that. It was like thousands and thousands. And that's on a daily basis."

If there are more than about twenty deer per square mile, Rawinski says, they'll quickly deplete the surrounding shrubs. The net result is the forest that many of us grew up with in the Midwest and East Coast. A forest of trees, leaves, and not much else, beyond the tougher, untasty species like mountain laurel. "Most people have grown up not knowing what a healthy forest should look like," Rawinski adds.

Reduce the deer, though, and you begin to learn. When Thom Almendinger became the director of natural resources and agroecology at the Duke Farms Foundation in Hillsborough, New Jersey, deer were suffering on the property. "We had deer that were succumbing to malnutrition," he says. "One winter I remember watching them eat blue spruce and boxwood, you know, just to get stuff in their stomach." Overpopulated deer can also pass on diseases more easily among each other, including epizootic hemorrhagic disease (EHD), chronic wasting disease (CWD)—even COVID-19.

The forest was suffering too. Almendinger describes it as "geriatric," all adult trees, all the same age, with no young growth coming in to replace them. Without undergrowth and saplings, birds and mammals were also disappearing. "There's a whole suite of species that don't breed in the trees, they breed in the understory, if you don't have an understory they disappear."

Almendinger and his team at the Duke Farms hired a company with sharpshooters in 2004. They removed 220 deer in three nights on a single square mile, he says. Then he put up a fence to enclose the now deer-free square mile of the property. On the rest of the property, seventy-five volunteer hunters harvested an additional 395 deer. After the initial cull, deer outside the fence are kept lower with hunting and deer drives, human chains of people walking in lines through the forest, herding the deer ahead of them toward sharpshooters in the trees.

Since 2004, the forest has come racing back. In a 2020 study, Almendinger showed that tree saplings were twelve times more abundant inside the fence and five times more abundant outside.

Native species grew back (though invasives did well too). "We're in the middle of doing research, but a lot of small mammals are returning," he says. "You know, we didn't have rabbits. We didn't have chipmunks. They had no cover in the past and the predators went wild." Now, they have understory. And the small mammals are back.

EMOTIONS AROUND WHITE-TAILED deer run high. There's my Aunt Carrie and her dogs, and her friend and her vats of deer corn. In Massachusetts, there's the threat of Lyme disease and the role deer populations play as host to ticks.

Lyme disease—caused by bacteria in the *Borrelia burgdorferi* species complex (oh, yes, bacteria get species complexes because they are that weird)—is permanently connected with ticks in people's minds. It should be. The tick (specifically the black-legged tick, *Ixodes scapularis*, sometimes called the deer tick, or *Ixodes pacificus* on the West Coast) harbors the bacteria. Ticks are bloodsuckers and they aren't picky. Birds, squirrels, deer, lizards. If it's got blood, it'll do.

When the tick (usually in the nymph stage, so small it could be mistaken for a poppy seed or an unfortunately large blackhead) bites a person and hangs on for more than thirty-six hours, the bacteria can transmit. The net result is someone with Lyme disease—symptoms include a fever, vicious headaches, aching joints, and weird bull's-eye-shaped rashes (believe me, I've had it). Some people—like me—feel better after a few unhappy weeks on strong antibiotics. Others suffer for months or years. While only about thirty thousand cases get reported to the CDC every year, up to 476,000 people might get infected yearly in the United States alone.

For most of the nineteenth and twentieth centuries Lyme disease wasn't on anyone's mind. Then, in 1975, thirty-nine children and twelve adults turned up with an odd disease around Lyme, Connecticut, and in 1977, the disease began to be known as "Lyme arthritis." Up until then, ticks weren't much of a worry on the East Coast. Because so much of the Northeast was cleared of forest and hunted out of deer, ticks—which needed those deer—were limited. Only after we rebuilt the forests—and the delicious, delicious forest edges—did deer, and the tick, come back.

But the relationship between deer, Lyme, and humans is trickier than it looks. Deer are a pretty poor host for the bacteria, and the ticks pick up the infection from other species. "The white-footed mouse (*Peromyscus leucopus*) is the one that has the highest reservoir competency," says Samniqueka Halsey, an applied computational ecologist at the University of Missouri. "It is really good at transmitting bacteria, and it has ticks, and it's everywhere."

The mouse might be the best bacterial reservoir, but deer are the best reservoir for ticks. The average number of adult ticks on a deer varies according to which study you read. A 1992 Illinois study counted 7.1 ticks on average per buck and 3.6 per doe, while a 2012 Pennsylvania study averaged 12.7 ticks per deer. Those are just the adults. Looking at the teeny-tiny larvae (even smaller than the nymphs that usually get humans), scientists have found about 555 larval ticks per deer.

If we have a lot of forest edges, a lot of deer, a lot of mice, and a lot of ticks, that leaves us with a few options. Efforts at Lyme management can mean trimming back undergrowth and doing controlled burns, preventing the tall grass and plants that ticks hang out on thirstily waiting for your bare ankles. Managers can install bait boxes and salt licks that give deer or mice a dose of antitick medication (called an acaricide). Wildlife managers can also educate the public on preventatives like always hiking in long pants, wearing bug spray, and doing tick checks. There was even a Lyme vaccine approved in 1998, though it got removed from the shelves after fears of side effects and low demand. There's now even hope for another. Good luck.

And of course, they can remove the deer. "It's a two-point

process" to getting rid of ticks, says Sam Telford III, a medical entomologist at the Cummings School of Veterinary Medicine at Tufts University. Scientists could attack the animal that feeds ticks—allowing them to breed—or attack the animal that infects them with the Lyme bacteria. "And I think that it's far more effective to try to attack making the tick as opposed to infecting the tick."

Starving ticks aren't going to devote energy to making more ticks. That means getting rid of the deer. "I'm the nation's biggest advocate for killing deer to try to reduce the risk of Lyme disease because of work that I did during my doctoral studies," Telford says. "We did the first and definitive experiment where you reduce the deer and watch the ticks correspondingly diminish."

The connection between ticks, Lyme disease, and deer doesn't just increase people's distaste for ticks. In a study of white-tailed deer management in the Blue Hills Reservation in Massachusetts, Anne Short Gianotti, who studies conservation and environmental policy at Boston University, got to examine in real time how populations of deer—and increases in Lyme disease—affected people's attitudes toward their local Bambi.

The Blue Hills Reservation is a woody spot just outside downtown Boston, surrounded by suburbs. Since the mid- to late 1990s, of course, it's been a great spot for deer. In 2013, a survey estimated eighty-five deer per square mile in the Reservation. In 2015, the Massachusetts Department of Conservation and Recreation organized a controlled hunt to bring the numbers down. The deer were overabundant. But Short Gianotti questions what "overabundance" really means. "The term itself has a function of creating the idea that the species is a pest in some way and deserves to be managed. It's, of course, a human subjective decision about what counts as overabundant."

Conservationists might say "overabundant" is when the deer are causing environmental harm, as Almendinger did. "I'm looking at ecosystem function. How is this system working for wildlife, that's my goal," he says. He wants to see native plants flourish, native birds and mammals come back.

That is also a human want—based on science, yes, but human nonetheless. The idea of deer overabundance, Short Gianotti points out, "is a human term that reflects human decisions about what landscapes should look like, what deer populations should look like, what deer suffering should look like, what human-deer interactions look like."

In the Blue Hills Reservation, for example, people came out against the deer hunts, but people came out in support of them too. One leader in the community, anonymized as Linda, ended up in favor of hunts after getting Lyme disease. After asking around, she realized that at least eighteen other people in her neighborhood had gotten Lyme disease as well, some several times. Short Gianotti noted that in talking to Linda, there was "something that approaches desperation in feeling like the situation just really, really needs to change." At the time, the solution seemed to be to get rid of the deer.

The problem is that when landscapes are about what we "want," there's lots of room to disagree. One side might say, for example, that the Blue Hills Reservation is a natural ecosystem, and to keep that ecosystem the way it evolved, the deer hunt needs to happen. The other side might say, yes, we have this natural ecosystem, and that means humans need to stay out of managing it. "The idea that something is natural doesn't necessarily lead to one conclusion or another about what's possible and what should be done," Short Gianotti explains.

The two sides lined up. Pro-hunting included the Friends of the Blue Hills, anti-hunting included the confusingly named Friends of the Blue Hills Deer, and the Society for the Prevention of Cruelty to Animals, who wanted the deer populations controlled with chemical contraceptives—or not controlled at all.

Deer (and horse) populations can be managed with chemical birth control in some situations. Females can also be ovariectomized, and males can get vasectomies. Birth control of any type works best when the deer population is in a place where other deer are unlikely to enter. For example, male deer are getting vasectomies on Staten Island. Female deer are getting ovariectomies in Ohio. Deer have been sterilized on the campus of the National Institutes of Health. Both the NIH campus and the Ohio location are in places surrounded by fences. They aren't islands—but for the purposes of the deer, they might as well be. All are the work of the company White Buffalo, a nonprofit that works in ecology and management of wildlife. Some of that is deer birth control. Much, much more of it is sharpshooting and managed hunts.

The problem with birth control is that bucks get around. "One male deer can impregnate many does," explains Jason Boulanger, the president of White Buffalo. "You have to go in and get most if not all of the males." Conversely, you have to get most of the does too. Every deer you miss is a deer that breeds. Both surgeries are expensive. In many states it's still only permitted as research. Vasectomies can be done behind a bush, but every doe has to be darted, popped on a stretcher, taken to a special area prepped for surgery, and then released. Islands, Boulanger says, have a higher chance of success with fertility control. If you can't get an island, he says, you can make one with fences, keeping a population of deer isolated. Sometimes, Boulanger says, areas try to control deer in both ways, a managed hunt or cull along with fertility control.

Fertility control probably isn't going to help in the Blue Hills. As with the horses, the current situation in Massachusetts pleases no one. The hunt there continues, and the deer continue too. Currently, there are estimated to be about twenty-three deer per square mile, well above the targeted number of twelve. The plants in the Reservation still show low numbers of small woody plants and seedlings, indicating overbrowsing.

In her study on white-tailed deer in the Blue Hills, Short Gianotti and her coauthor argue that the deer became killable— they became defined as a population in proximity to humans and causing humans problems. The deer altered the ecosystem and increased human Lyme disease risk. They became a pest.

In white-tailed deer, it's easy to see how we create our own pests. We intentionally and unintentionally grew habitat. We built towns where no one hunts and got rid of every predator. We saw, and still see, deer as beautiful, quiet, gentle nature. The deer came where they were wanted. Then we decided this was not the landscape we wanted after all—and not the deer we wanted either. Some of that is realizing that deer suffer at very high populations, but a lot of our about-face comes from our increasing—and increasingly negative—interactions with those deer.

As the deer cause us more problems, our tolerance erodes. We don't want them here anymore, but we can't bring ourselves to kill what we've created. If there's anything here that needs to change, it's us. We need to understand that our development and gardening choices don't exist in a vacuum. We need to learn what kinds of habitat we offer—and for which animals. Or, if we don't want to change the places we have built, we might need to change what we think an animal's "natural" fate should be. We need to change our beliefs about whether the animals that live closest to us are truly wild, and if not, whether and how we intervene in their lives—and in their deaths.

10

A SLOTH
OF BEARS

t's an abnormally warm February day in the Berkshires in western Massachusetts, and I'm standing in a half-frozen, swampy pocket of land between farms and houses. There's a persistent growl of traffic in the distance. Buildings are visible through the trees, and an American black bear (*Ursus americanus*) is passed out at my feet.

She is flat on her front, stretched out where she finally collapsed when the drugs took hold. Originally, she had denned in a slide of bark surrounding a downed pine in a thicket of brush right behind a large and very occupied farmhouse (complete with large, barking farm dog). It wasn't much of a den, more a flat spot for her and her three yearling cubs to snooze away the rest of the winter. She's denned in this area since the scientists first collared her in 2018, sometimes in the woods around a cemetery, sometimes on the farmhouse property.

Dave Wattles, the black bear and furbearer biologist (if you ever doubt that the modern American parks and conservation movement came out of hunting culture, "furbearer," meaning anything with a sellable pelt, is a good clue) for the Massachusetts Division of Fisheries and Wildlife, tracked her down by following the ping of her GPS collar to this thicket. The technicians loaded up guns with darts full of a potent anesthetic and surrounded the den at a distance, concentrating in the area she was most likely to use for an escape. Wattles went in himself, along with a biologist for the district who carried a long pole, also loaded with a dart. The goal was to get close enough to stick the bear and give her a shot of the drugs before she could run.

But black bears around here do not hibernate so much as lightly snooze. This one roused quickly and ran, her three yearlings scattering in every direction. One of the younger technicians aimed and fired, hitting her square in the rump with a dart. The bear raced off, crossed a road, and passed out in the swamp.

Her ear tags read "190" and "191," but the wildlife biologists and technicians usually call her "Fairfield" after the area where they first trapped her and fitted her with a thick leather collar with a very expensive GPS transmitter. Now, they drag her onto a tarp and commence taking off her old collar and putting on a new one, checking her body condition as they do.

Her fur is warm and soft. The firm pads of her feet are the color and texture of old avocado skin. Her tongue lolls out and she snores a little. She smells doggish, like clean German shepherd. Fairfield is healthy and raising her second set of cubs since she's been collared.

She's been around this area a good while, notes Dave Donald, the owner of the farmhouse. He shows me video from 2020 of four tiny cubs—her previous litter—gamboling on his lawn. He's used to Fairfield, whom he calls Misha. Donald has lived in the area all his life and built the farmhouse himself. He feels comfortable living with bears. A few months ago, he says, the dog started barking at the door around 9:00 p.m. "I turned on the light and a bear stood up" on the porch, he says, looking him square in the eye through the front-door window. He figures maybe he spilled some birdseed. But he didn't panic. He just didn't go outside.

The yard in the front is full of bird feeders, which are quite busy on this warm day. Donald insists he brings in his bird feeders every night. Wattles eyes the feeders narrowly. "It's the bane of my existence," he says. Birdseed, you see, is the peanut butter of bear foods. A relatively small volume packs a big caloric punch. As we drive through Dalton on the way to the bear den, he barks, "There's a bird feeder right there," "There's a bird feeder," as we pass house after house. He also doesn't have a lot of patience for the current craze for backyard chickens. Both are irresistible opportunities for bears—and sources of conflict. Enough bird feeders gone, and someone will call Wattles. If a bear tries chicken, someone might reach for their shotgun first.

Fairfield begins to come around, shifting her head back and forth, and the scientists heave her into a net and haul her over to a new den they've built, where her cubs will find her. They settle her in, cover her carefully with branches, and give her an affectionate pat. A technician shakes off the tarp, where Fairfield had defecated as the drugs began to wear off.

The poop falls apart easily. It's full of birdseed.

Western Massachusetts is a good place for bears. "After deforestation and colonial settlement, they were really just relegated to a small corner of the Berkshires," explains Kathy Zeller, a spatial ecologist with the Aldo Leopold Wilderness Research Institute in Montana. Like deer and wolves, black bears were hunted and chased out of large portions of the East Coast.

Hunting restrictions and the regrowth of forests have benefited more than the deer. Over the past forty to fifty years, scientists—and homeowners—have been watching as black bears move back in, rolling east across the state of Massachusetts. "So in the seventies, when our research really started here in the state, there [were] estimated to be maybe one hundred bears in the northwest corner of the state. Now they're estimated to be about five thousand," Zeller says. "And they're getting closer and closer to Boston."

It's a slow stroll, really. The young males show up first, dispersing far and wide away from their mothers. Females tend to remain close to where they were born, developing partially overlapping territories with mom. But those territories, too, march east. And now, here we are in Pittsfield, looking at a bear who's only ever known a world with roads, cars, and plenty of birdseed. If a mother bear is denning in a suburb, Wattles says, "she's going to come out of her den in spring, there's not much food available, she's gonna go to backyards and feed at bird feeders." If she's got cubs, the cubs do the same, and raise their cubs to do it too. "From the age of nothing, those cubs are being taught those behaviors and those places to find food, and places they should spend time," he says.

Wattles and I drive up to a tree service company in Northampton. They'd called to ask about the bear denning on the property. We hustle over the frozen ruts to the back, an area with huge walls of cut trees and a giant pile of wood chips. Three men work a chain saw nearby. The noise is deafening. Less than ten yards away, the body of a dump truck has been set on some cinder blocks for the winter, a gray, silent hulk. Under it is a bear Wattles calls Deck and her one yearling. They seem totally impervious to the roar of the chain saw.

Deck is a big girl with a cinnamon nose and a history of denning under decks (hence her name) and by the pool in a nearby apartment complex. She gets a lot of birdseed and trash, but also dines on skunk cabbage and natural foods from the nearby wetlands. A bear's territory in a crowded area like this can be as small as forty square kilometers, Wattles explains. She could den in a quiet, more natural area. But her little one needs easy food, and the easy food is here.

When Wattles asks the employees about her, they say, oh yes, they saw her up on top of the hill on Monday. They aren't concerned. They just want to know when she moves on. They need to use the dump truck eventually.

Black bears prefer to den in trees when they can get them, explains Jennifer Strules, a graduate student in wildlife, fisheries, and conservation biology at North Carolina State University in Raleigh. When colonists came in and cut down most of the old growth on the East Coast, big trees with bear-size holes became hard to come by. Instead, the bears go to ground. They'll pick a spot that's fairly hidden—behind brush, under fallen trees. There, they might do a little digging, scuffle around, "like when dogs turn around to make a little nest for themselves," Strules explains (though she's guessing, she hasn't seen a bear den down in real life). Into their little bowl-like nest, they might drag some leaves or brush, or they just take it as is, as Fairfield tends to do. Then they settle in. Unless, of course, they happen to have picked a place that wasn't under trees at all, like under someone's deck, and that someone is not pleased to see them.

In and around the city of Asheville, black bears den under porches and buildings, or mere feet from highways. One found a home in an abandoned car shop, where a section of the concrete foundation had been cut down into the ground to allow mechanics to get underneath cars. With the mechanics long gone and brambles grown over the cavity, a black bear moved in. "This little underground hidey hole was there and that's where she was," Strules says. "She had two cubs."

The cars and oil changes might have been gone from the area, but people were still around. A retail store was nearby, and "there was an active homeless camp maybe a hundred fifty feet away," Strules says. There's this idea that "these animals that live in the hinterlands, you know, have to be far away from people in order to survive and reproduce," she notes. "If you visit our den sites . . . that view is going to be challenged pretty quickly."

Hunting and habitat destruction—including the loss of the American chestnut—shoved the black bear out of most of the East Coast, and conservationists have been trying to bring them back. Scientists even reintroduced them around Tennessee, southwestern Virginia, and Kentucky. In North Carolina, protected bear sanctuaries in places like Great Smoky Mountains National Park helped populations recover.

The bear comeback is one of North Carolina's biggest wildlife success stories, says Colleen Olfenbuttel, the black bear and fur-bearer biologist for the North Carolina Wildlife Resources Commission. The bears came back and they are thriving, a sign of an ecosystem that can support them. It's thrilling to see such charismatic megafauna in the relatively tame eastern woodlands. "There's a lot of doom and gloom with many wildlife species," she says. "But this is one that is a success story that we should all be proud of."

However, when bears are living around humans for the dumpsters, the bird feeders, and the dog food that residents supply, Olfenbuttel worries that success will lead to new failures. Thrilled people record cell phone videos of black bears in pools, on the lawn, in trees. These are positive interactions. But as black bears get more comfortable, interactions get negative. They become conflict. Bears get too close. They're "getting rewarded for being near people. And that's not the lesson we want to teach them," she says. The bears get photographed and fed instead of punished, and keep coming back. "That's probably one of my biggest concerns is that as conflicts continue . . . tolerance for bears on the landscape will decline."

Wattles worries about it too, especially as bears get closer and closer to the suburbs of Boston. Out in western Massachusetts, people are used to bears. In the city? Yikes. "The idea of a bear wandering through, that is a push-the-panic-button moment," he says. "And things can escalate in a real hurry out there, just because of the density of high-speed roads, the density of development."

Too much conflict, and black bears will transform in people's eyes from inspiring signs of wilderness to a giant raccoon. "I've seen how, you know, tolerance has declined for these species because of their abundance," Olfenbuttel says. "People associate [raccoons] with diseases and causing conflicts." Bears, too, she worries, could end up the next trash panda—the next pest.

LET'S BE CLEAR, bears love what we love—snacks. When conflict between humans and black bears occurs, it is often because a bear is tempted by human food, or something that passes for it. Bears are not picky eaters. Crops are delicious, so are the tubers and nuts and berries of the forest. But why bother with health food? They'll eat garbage or compost. Fast-food dumpsters? Sign them up. Dog food or birdseed? Count them in. Salmon is tasty, and so are chickens and eggs. They'll lick the burned bits off your grill and chomp on your Halloween pumpkin. I know some guys who got a visit from a bear in the night because someone had toothpaste in the tent. It's all fair game to a bear.

Part of the reason bears aren't picky is because they have to eat so much. In some areas, black bears can fully hibernate in the winter, not eating, drinking, pooping, or peeing for up to seven and a half months. Even if they don't fully hibernate—just move less and sleep more—winter is a poor time to find food. So in the summer and fall, forget your two thousand calories a day—bears need twenty thousand. They're unique in that they take food cravings to the next level, explains Stewart Breck, a wildlife biologist at the USDA National Wildlife Research Center in Fort Collins, Colorado. In late summer and fall, they'll forage and feed up to twenty hours a day.

Wattles talks about a bear that splits its time between western Massachusetts and New York. "Every single fall, she goes down into New York and just absolutely hammers cornfields." She eats so much, he says, her crop circle–like holes can be seen on Google

Earth. A bear in the fall is a large, hairy, food-seeking missile, with an incredible nose and memory for where it's seen food before and where it's likely to get it again.

In a now famous study in Yosemite National Park, Breck and his colleagues analyzed the cars that bears broke into. Between 2001 and 2007, 908 vehicles got a bear visit. When the scientists looked at which types of car were available, they found that of the cars in the park between 2004 and 2005, bears had a major preference: minivans. The bears headed for a minivan four times more often than would be expected by chance.

Why are bears into minivans like they just hit thirty-five and had a third kid? Breck and his colleagues weren't able to find out for certain, but he's pretty sure the answer is the kids. Minivans are more likely to be driven by families—which means back seats with a fine layer of Cheerio dust and smearings of Go-Gurt. To the sensitive nose of a bear, that smells like jackpot. "It's not that I have hard evidence that way," Breck says. "But it's very likely they're selecting vehicles based on probability of smell or finding smells, and thinking, 'All right, there's high probability of food near.'"

That was in a national park, where people expect bears. They know they'll have to guard their behavior and might actually read posted signage about wildlife and follow its guidelines—or at least, they think they will. They know they're in a special place—the wilderness, a place where you expect to find lions and tigers and bears. But at home, in the suburbs, bears are less expected, and people definitely aren't as excited to see them.

There's good living to be had near people. In North Carolina, it's so good, in fact, that bears in urban areas are outgrowing and outreproducing bears in rural areas. "We have a saying that when bears leave Asheville, they die," says Chris DePerno, a wildlife ecologist at North Carolina State University in Raleigh. In a study examining forty years of yearling bears, he showed that young bears living in Asheville were much larger than rural bears living in forests nearby.

DePerno and his colleagues visited urban and rural bear dens and lured bears into live traps with the bear scientist's mainstay: day-old supermarket pastries. "We work very well with the local supermarkets," he says. "We got the stuff for free."

Normally, female black bears have their first litter of cubs when they're about four. But in the urban environment of Asheville, they start family life much earlier. In DePerno's study, urban bears started breeding much younger than rural ones—58 percent of the urban bears were having cubs at two years old, while none of the rural bears bred at that age.

To live successfully with people, bears—like mice or raccoons—end up cultivating a new set of skills. In a 2019 study in western and central Massachusetts, Zeller and Wattles collared seventy-six female black bears and tracked their movements via GPS between 2012 and 2017, following bears in an area that stretched from Albany, New York, to Dover, New Hampshire, and even as far south as Rhode Island.

A human landscape is what some ecologists call a "landscape of fear" for a bear. Humans are scary—even if you're the size of a bear, Zeller explains. We have cars and guns and traps. But we've also got bird feeders and dumpsters. So a bear has to weigh the risk of encountering humans with the reward of the food we generously provide.

To handle the risk and get the reward, bears change their behavior. In the wild, bears tend to be most active around dawn and dusk, bedding down for the night from about 11:00 p.m. to 3:00 a.m. Bears living near people, though—whether it's campgrounds or suburbs—will take up the night shift and reduce their activity during the day. Zeller and Wattles showed that bears haunted roads and housing areas at night, especially during the spring and fall. There was a better chance of getting a good meal and passing unnoticed—at least until the neighbors all invest in camera doorbells. Over time, bears living near people learn to be active at night but also let their guard down a little, avoiding paved areas less than their wild counterparts.

They even learn to cross the street. Another study that paired GPS collars with bear heart monitors showed that bears crossed an average of one road a day, and every time they did, their heart rates increased (sadly, the monitors could not detect whether the bears looked both ways first). Not only do bears know where to find human food, they also know full well which parts of human-based living are the most dangerous. But they don't do it successfully forever. As we drive around Northampton, Wattles shows me den after den—in railroad culverts, the bushes off the expressway, behind a lawn-care company. Most of his bear stories end with the bear dying—usually while attempting to cross Interstate 91 or one of the other big roads nearby. Several of the bears he's collared, he notes, have limps from near escapes.

The danger is spreading. People are moving more and more to places with good bear habitats—like western Colorado—says Heather Johnson, a research wildlife biologist with the U.S. Geological Survey in Alaska. Populations of bears have also increased in places like Massachusetts, North Carolina, and elsewhere over the past few decades, in part due to reforestation (similarly to the deer), but also due to changes in hunting.

Johnson and her colleagues have shown that black bears in places like western Colorado and Nevada tend to head into human-dominated areas when the pickings are slim in their natural habitat. The issue is something called low "mast years." Food in the forest is divided into hard and soft mast. Soft mast is stuff like chokecherries, blueberries, apples, and so on. Hard mast is grub like acorns and pine nuts. In another study collaring 109 bears (sometimes it can seem as though half the bears in the country must be wandering around in the woods with GPS jewelry) from around Aspen, Tahoe, and Durango, Johnson and her colleagues showed that bears were especially likely to wander into town when mast years were poor.

The lure of the suburbs might be especially tempting due to climate change. Droughts, Johnson explains, contribute to mast failures. So do those random late freezes in the spring and early summer. Both of those weather events are likely to occur more frequently with climate change. "We should expect more natural food failures in the future," she says. And when natural food fails, a bear will need to find something else to eat.

Warmer weather also shortens a bear's long winter's nap. Johnson and her colleagues have shown that if fall in Colorado is 1 degree Celsius warmer on average, a bear's hibernation is 2.3 days shorter. The bears also slept less when there was more food to be had—both human and natural food. Fairfield and Deck, for example, didn't need to wait for skunk cabbage to come up in the spring. Bird feeders were full all winter, and humans never stopped putting out trash. Better weather and big dumpsters? Why snooze on that?

We think of bears coming too close as dangerous for us. And it can be. But it's usually worse for the bears, Johnson explains. "The more time they interact with people, it's riskier for them, the more chance [they] have of dying."

Another study from Johnson's colleague, wildlife ecologist Rebecca Kirby, used blood and hair samples from 296 Colorado bears to find out what they were eating. She checked the samples for carbon 13, a form of carbon that is especially enriched in the things that mark the American diet above all others—corn and cane sugar. Bear hair and blood enriched in carbon 13 means a bear dining on corn, Coca-Cola, and the myriad of other things we shove corn and sugar into—from Pop-Tart fillings to the livestock that fill up on the corn we feed them.

Kirby showed that for every 1 percent increase in carbon 13— for every percent increase in corn dogs or ice cream—a bear's odds of being labeled a "nuisance" bear shot up by 60 percent. Bears are super cute on social media, until they're not. Until your dog goes barking up to one and the bear takes up the challenge. Until your new backyard chickens end up as a bear's KFC. Until a bear real-izes you keep food in your garage.

A corn-fed bear is also not necessarily the healthiest bear. Kirby and Johnson have both shown that bears dining on human food will hibernate less. But hibernation for a black bear is more than saving energy when food is thin. It's also about staying youthful. As cells age, their telomeres—long repeating sequences of DNA on the ends of chromosomes—get trimmed. Hibernation slows that trimming, suggesting hibernating could be a strategy toward living longer.

But a human food–fed bear that's not hibernating as much is missing out on this beauty rest. Kirby and her colleagues have found that bears hibernating less also had shorter telomeres—a sign their cells could be aging more rapidly.

The net result is that in Colorado, human-dominated areas are what's known as a population sink for black bears—which means more bears go in than come out alive. In another study of the bears in Durango, Johnson and her colleagues showed that a poor natural food year resulted in a 57 percent decrease in the number of female bears. It wasn't that bears were dying of hunger. The decrease went along with a major increase in human-caused death. The bears had been driven by poor natural foods and lured by our rich ones. They braved human-dominated areas, and they died there, most often "harvested" (that's hunted), crushed against the bumper of a car, or removed by wildlife management—for being a pest.

BY NOW, WE know the ideal way to stop animals from coming too close to humans. No food? No bears. If people locked up their trash, put away their bird feeders, cleaned their grills, took in their dog food, and kept a strong eye on their fruit trees, bears would find fewer reasons to come calling.

Telling people what to do is one thing. Getting them to do it is another, as Johnson and her colleague Stacy Lischka found out. They started out with a goal—bear-proof a city. "The assumption underlying this project is that if we reduce conflicts, we would increase people's tolerance for bears," says Lischka, principal consultant for Social Ecological Solutions in Fort Collins. It's not a wild assumption to make, she notes. In fact, the idea that if you reduce conflicts people will love bears is widespread in bear management across the United States.

The scientists decided to bear-proof the city of Durango, Colorado, which during the period of their study (2011–2016) had about eighteen thousand people and a lot of human-bear conflict. Durango is in the Front Range, in an area full of oak and scrub habitat—bear country. The human population had also shot up by 67 percent in the previous forty years, and the small city had become a hot spot in the state for bear roadkills.

Bear-proofing the whole city was outside the budget, and the scientists needed areas that could serve as a control. They pinpointed places on both the north and south sides of Durango, each with assigned control and experimental sections. In the control sections, they changed nothing. In the experimental sections, Johnson, Lischka, and their colleagues went all out. Beginning in 2013, they handed out more than a thousand bear-resistant trash cans—made of thick, heavy plastic with not one, but two latches that needed to be locked. The free trash cans were no small feat, as a single can is easily more than two hundred dollars. By the time the team was done, more than 95 percent of homes in their treatment areas had the bear-resistant cans.

The scientists didn't stop there. They went around to houses in the neighborhoods, reminding people to lock their cans. They asked them to clean up other bear attractants like bird feeders and pet food. They handed out flyers like missionaries of bear tolerance. Do you have a moment to talk about bears?

And the scientists spied on the residents. Every summer, Johnson says, they hired someone whose only job was to cruise neighborhoods at five in the morning, checking people's trash cans. If someone's can was open, they got a big sticker on the can, letting them know they were in violation of the city ordinance. They also got a big orange door hanger as an extra dose of social shame. Free can? Carrot. Big orange violation sign? Stick.

Someone driving slowly down the street in the dawn light staring intently at your garbage is a little unusual. "I will say we definitely got the police called on us," Lischka says. But the trash spy also allowed Johnson and Lischka to check reality against what the people in their study said they would do. After all, they could (and did) send out mail surveys to ask people if they were using the cans. "But being the good scientists that we are, we also thought we better check. That was one of the most important decisions that we made in the whole project."

It turns out that you can lead humans to solutions for human-bear conflict, but you cannot make them lock up their trash. People did use the cans. But many residents just used them like regular garbage cans—which meant they were perfectly open to bears. They'd leave the lid open, not lock it, or overfill it and put a nice big bag of trash next to it on the ground.

"We assumed when we started the project that . . . the impediment to people using bear-proof garbage cans is the fact that they cost two hundred and fifty bucks apiece, and who's going to spend two hundred and fifty dollars on a garbage can," Lischka says. "We thought we'd have like ninety-eight percent of people using them because we gave everyone a can." After five years of study, it was clear that money was only half the battle.

When the scientists asked the residents if they were using the cans after the first year, more than 70 percent of the participants stated that yes, of course they were using them correctly. But the bear-proofing spies found out only half actually did. By 2016, when everyone had a can, only 24 percent of residents used it properly all the time. Another 34 percent of the residents never used them correctly. These somewhat depressing numbers were probably what Johnson politely calls "behavioral intention." It's not that they were lying, exactly, it's just that their intentions didn't line up with what they did. How you think you're going to use the trash can is not how you're going to use the trash can.

It wasn't all hopeless. In fact, human-bear conflict declined almost immediately after the bear-resistant cans came in. Bears got into the trash 60 percent less in the treatment areas than in the control, and people locked up their trash 39 percent more. On some blocks, people were very bear aware, and conflict dropped. And the cans and flyers and reminders did make people feel that bear management could work, and made them worry less about bears roaming the neighborhoods in search of trash.

The people of Durango don't want bear conflict—no one does. It's just hard to remember your good intentions when it's 10:00 p.m. and the trash needs to go out and the dishes aren't done and the dog needs to be walked and one kid won't do their homework and it's freezing and now you need to take in the bird feeder too? When good intentions are forgotten once, it's even easier to forget them a next time, and so on. As anyone who's ever tried to keep a New Year's resolution knows, changing habits is hard.

THE GREAT BEAR RAINFOREST in British Columbia is, as you might imagine, stuffed with bears—not just the American black bear but also grizzly bears (*Ursus arctos*). People live there too, including people of the Kitasoo/Xai'xais First Nation.

They have a long history of living with bears, explains Douglass Neasloss, stewardship director and chief councillor of the Kitasoo/Xai'xais. Once, he says, he went to a conference on how to bring Western scientific methods and Indigenous knowledge together. "A First Nations lady, she said, 'What is science? Knowledge and observation over time? First Nations, they have ten thousand years plus of observation over time.'" Neasloss and many of his colleagues have been trying to work with scientists to make sure some of that traditional ecological knowledge (TEK) and the sources of that knowledge, like Neasloss and his community, are included, and respected, in scientific research.

A lot of the TEK that Neasloss references is passed on by experience and stories. "Our laws come from our stories," he explains. And all of those stories have wildlife in them, as symbols, and also behaving as themselves, interwoven in Kitasoo/Xai'xais culture. "We're always taught to have respect."

One of the most important tenets Neasloss says he remembers from stories is this: Don't mock the wildlife. He says he can't help but notice the difference from the stories that people growing up in European-dominated societies tell—stories like Little Red Riding Hood, Goldilocks, the Three Little Pigs. The potential predators are all bad in those stories. Wolves are slobbering. Bears are . . . well, Goldilocks ran, no matter how the bears behaved. In the stories he heard growing up, Neasloss says, animals were sometimes to be feared, but they also had their own dignity.

As an example, Neasloss told me a story about a woman and a bear. The woman's father had killed a bird, and because of this, spring never came to his home. Frustrated, the father had moved his family to an inlet—where spring had come—to collect elderberries. "It started to get a bit dark," Neasloss says. "The mother was up there with her kids. And as they were coming down, one of the daughters, the basket broke on her and all the berries dumped on the ground." The daughter told her family to go ahead, and picked up the berries. The basket broke again. While picking up the berries again, the woman stepped in a nice big pile of bear scat. She was completely grossed out and didn't bother to hide how disgusted she was. When she tried to move on, and the basket broke a third time, Neasloss says, "there was a voice behind her. It said, 'is your scat better than mine?'" The woman whipped around and there was a man—but not a fully human one. Instead, she faced a bear that could take off his skin and look like a man.

The girl lied. Yes, she said, my scat is better than yours. "She took off her copper bracelet," Neasloss says. "And she pretended to use the washroom." As she stood up, she dropped the bracelet, trying to show the bear that she crapped copper. The man-bear saw her lie, and took her away to his cave in the mountains, where they had several children. Finally, one of her brothers found her, as she threw snowballs down the mountain to give away her location. But her children—which could be bears, or could change their skin for those of men—left their mark on the mountains. "And we have some rock paintings up here," Neasloss says. "And these rock paintings are supposed to be the places where the children go." They left their handprints—and their pawprints—on the stones.

"The whole point of that is number one . . . we're not supposed to make fun of wildlife," Neasloss says. Don't assume that the bear isn't as smart as you are. "And then the other part is just to mark the close relationship with bears." Bears and people aren't so different. Coexistence with bears is very important to the Kitasoo/Xai'xais, and so is humility.

The Kitasoo/Xai'xais have a strong sense of the benefit that bears provide in their community, Neasloss notes, far beyond the benefits of tourism (though there's plenty of that). The community has a cannery and two rivers where salmon spawn. Where there are salmon, there are bears, both black and brown. Those bears take salmon from the river and transport it to the surrounding forest, in the form of fish heads and bones, but also in the form of bear bowel movements. Those nutrients fertilize the soil. The bears do digging in the estuary, Neasloss says. "You take away the salmon and you lose the bear, you lose the bear and you lose your forest," he notes. "[We] kind of look at things from more of a holistic approach."

A few years back, Neasloss says, his community, about an

eighty-kilometer ferry ride north of Bella Bella in the Great Bear Rainforest, did have some problems with bears. A few people had left trash out, and there were some fruit trees. Bears began entering the village. Someone ended up calling in the government, and bears ended up shot. (More than five hundred black bears were killed in British Columbia in 2021 for getting too close to people.)

The community was horrified, as they wanted the bears out, not shot. "So I phoned them," Neasloss says. "And I said, 'Can you guys not shoot these bears? Can we look at some other options?' And they said, 'No, our only mandate is to shoot.' So we basically said, well, you're not welcome in the community again, and we'll develop our own program."

The community ran a study, asking everyone what they wanted to do about bears in the area. "It was like something like ninety-five or ninety-seven percent, you know, said we should not be harming, we've got to learn to coexist, we've got to live with one another," Neasloss says. "It really just validated what we needed to do."

Now, the village has no more fruit trees. Every house has a large bear bin in the yard, where they dispose of any waste. There's also a live-bear trap that can be used to relocate bears. Just in case. "We'll do everything else before," Neasloss says. "We don't want to kill them." People slip up sometimes, he notes. Of course they do. But most of the time, there's a strong sense that this is just what you do when you live with bears. Now, the Kitasoo/Xai'xais are part of a larger effort to buy up hunting licenses in the surrounding area. They will live with the bears, and they won't condone other people shooting them, either.

Neasloss's community did not try to change the bears. They changed themselves. And they're not alone. Other towns such as Virginia City in Montana are making huge efforts to stay bear aware. They're forming groups to make sure every bit of fruit from fruit trees gets cleaned up before it attracts bears. They're using bear-resistant trash cans and using them properly. The difference between a community that lets bears forage in their garbage then punishes the bear, and a community that is bear resistant, isn't ability or a spiritual connection to the landscape. It's desire—the desire of the community as a whole to learn what they need to do, and to integrate bears and bear-safe practices into the community. You obey a stop sign to live with people. You take in your bird feeder to live with bears.

That desire doesn't come from simply not wanting to have conflict, Lischka notes. It comes from people seeing bears as a positive thing—not a negative thing you need to protect yourself from. To reduce human-bear conflict, she says, we have to make sure we don't view bears as a pest. "If you really want to change tolerance, you actually have to start to think about how you communicate . . . and how you might even try to attempt to manage for the positive interactions," she says.

When Lischka and Johnson carried out a survey of Durango residents' views of bears, they found that people's negative bear experiences—bears in the trash or damaging their property—didn't really predict their tolerance for bears. Instead, it was the positive experiences they had with the animals. "They were things like, 'I enjoy seeing bears when I was hiking in the wilderness; I enjoy seeing bears in my neighborhood. Knowing that bears are here, and have a viable population is important to me, and you know, a signal of the quality of the ecosystem,'" Lischka tells me. People who believed in bear benefits were more likely to accept everything that living with them entailed.

"We can't just always talk about bears as a problem," she says. People aren't going to have the desire to change their ways if they're only managing a nuisance. If they're managing a natural neighbor, one whose well-being they care about, on the other hand, they might take more action.

And people will need to be tolerant, because bears—and humans—aren't giving any ground. "We will likely never get to a point where we have zero conflict" in many places, Lischka says.

"Tahoe is one of those places, Aspen is one of those places . . . the quality of bear habitat is too good. . . . So we have to get to a point where we can figure out a way to manage to the lowest possible level of conflict." Conflict will still happen, because people and bears make mistakes. "But we have to get to a point where people are okay with that."

If we don't—if we let ourselves think only the worst of our bear neighbors—we won't think of them as neighbors at all. Instead, we'll think of the woods we move into as belonging to us, and bears as nuisances that failed to learn they weren't wanted. We'll manage them like we manage any other interloper. We'll grow annoyed, fearful, and angry. And bears will become pests.

11

A PEST BY ANY OTHER NAME

One weeknight in August 2021, my friends Emma and Paul (names have been changed to protect the slightly embarrassed) had put their daughter to bed in their suburban Maryland home. They were sitting in the living room, watching *Star Trek: Deep Space Nine* reruns.

"I heard a rattling sound," Emma says. They'd been having mouse problems recently, so she thought perhaps it was another mouse that had ended up in one of their traps. But the sound kept getting louder. "I was thinking to myself 'geez, did we get a rat or something?'"

Then Paul turned around and yelled, "What the fuck is that?!"

I found out fairly quickly what was going on, as Emma posted a frantic message in one of our group chats. It was a photo, showing a corner of her living room, with bookshelves and board games piled on top. There appeared to be a stuffed animal wedged in the very top corner, up against the ceiling, behind a copy of the game Wingspan.

It was not a stuffed animal. "We turn around and there's this raccoon on the top of our bookshelf where there normally is not a raccoon," Paul says. The animal had wedged its nose as far into the corner as it could, and clearly was trying very hard not to be seen. It had most definitely failed.

Complete panic ensued. "I go into my fight-or-flight mode, which for me is flight," Emma says. She grabbed her phone, snapped a photo, ran into her daughter's room, and locked the door. Paul pointed out that, despite her fight-or-flight mode, she did manage to get a clear picture. The modern panic response: fight, flight, or photo.

Frantically Googling animal control, Emma realized their offices were closed. She called 911. The emergency dispatcher told her that they should just wait until the raccoon left. That's when Emma started messaging all her family and friends who might be awake.

Meanwhile, Paul, a martial artist, grabbed a sword off the wall and stood in the living room doorway. "This feels like the bad kind of sword-owner thing," Emma says. "Skewering a wildlife." But neither of them was thinking especially clearly. "I wasn't sure if it was going to stay there, or if it was going to, you know, run out of the house," Paul explains. "I wasn't sure if it was possibly rabid."

Luckily, Emma's friends (I was one) were up. We told her to go back, open the door from the living room to the porch, put a can of something smelly (preferably tuna) outside to guide the way, turn off the lights, and leave the room. By this point, Paul had already tried donning work gloves and boots and poking at the raccoon with a hiking stick, trying to make it leave. "It ignores it," he recalls. "Except when I poke it a little harder, near its head, it bites the stick until I take it away."

After some more back-and-forth over group texts, and trials

with crackers, sliced ham, and finally tuna, the lights were all off, the glass door to the porch was open, and Paul took up vigil by the dining room doorway. The only light was the glow of their Nest home security system, which lit up briefly every time it sensed the raccoon moving in the room. "I saw the Nest glow," he says. "And the glowing was this time accompanied by some knocking and scrambling sounds that sounded really promising to me. And then eventually, it just flashed across the floor out of that door and it did not even stop to consider any of the food."

The whole scene, Paul says, felt like it took forever to unfold. "We waited for a long time. It was almost midnight by the time it finally left," he recalls. I checked the time stamps from the group chat. Emma's first panicked message arrived at 10:02 p.m. Her relieved "It's gone!" at 11:00. The entire episode had taken less than an hour. I can only imagine what it felt like to the raccoon.

If this seems like a bit of an overreaction, well, yes. But most people don't have any up-close and personal experience with a raccoon, let alone a raccoon inside their house. "My only experience with raccoons is them robbing our bird feeders, you know, tearing the bird feeders off of our deck and dragging them into the woods," Paul says.

The pair spent the next few days disinfecting every object in the living room, and then finding and boarding up any holes where the raccoon might have entered the house.

Paul knows that the sword reaction was absurd. "In hindsight, the animal was completely terrified. If anything, we should feel pity for it," he says. "But that is not the first response to what feels like a threat in the moment." His response was so intense, he says, because he felt like his space had been violated. "It's my house," he explains. "It doesn't belong in my house where, you know, we're supposed to be secure from wild animals."

In every pest story, the same themes come together. An animal arrives where it's unexpected, because it is sliding into a niche—

maybe a physical niche on top of the bookshelf. Perhaps we have given it a new habitat, or new opportunities like bird feeders. Or maybe we blundered into its habitat with our leafy subdivision. When my friends confronted that raccoon, they felt vulnerable and powerless. They felt disgust and fear, and were immediately worried about rabies—which raccoons can spread—and the pain of being bitten. They confronted their own belief that wildlife in their neighborhood was supposed to be a good thing, and found it clashing with the belief that their home should be an escape from nature. They saw how animals work their way into a niche, and how the habitat changes that produced their subdivision contributed to it. Some of these things they saw at once, some the next day, and some even later. But at the time, they all boiled down to one thing: Get. It. Out.

If we insist on constantly fighting a war against the pest, I am convinced it's not a war that we can win. Any expectations we have to exterminate are years in the future and probably limited in scope. In fact, the National Association of Exterminators and Fumigators (now the National Pest Management Association) decided to make the switch from "exterminator" to "pest control operator" in 1936. Pest control operator doesn't have the same ring, sure. But at least control was in reach. Extermination never was.

Yes, we can poison and trap until islands have no more rats or cats or mice. We can build bigger fences or develop new deadly diseases for rabbits. We can try new genetic techniques to render mice infertile and put contraceptives in pigeon feed. For every one we kill, another will step up to take its place. Maybe not a rat. Maybe a squirrel, or a raccoon. Maybe a deer, a gull, a pig. A bear, a turkey, a monkey, a crow.

As long as we continue to live the way we do, creating new spaces and new trash, bringing in new, exotic pets, moving into wild spaces, or clearing ones we don't value—animals that come to take advantage of us will always be in our way. They will always bother us. There will always be pests.

The animals in these chapters are just examples. In reality, many other animals could have stood in their place. Starlings, crows, gulls. Feral hogs, coqui frogs. So very many monkeys. That's not even counting the stinkbugs, centipedes, yellow jackets, and other invertebrates. Raccoons could have stood in for coyotes, monkeys for elephants, feral hogs for snakes. All of them animals that refuse to stay where we think they belong. Anthropologist Mary Douglas famously described dirt as "matter out of place." Pests can be seen in a similar way—as animals out of place.

But who determines what it means to be out of place? Humans do, of course. The feeling of dominion is a Judeo-Christian way of thinking, says Neil Patterson, Jr., a member of the White Bear Clan of the Tuscarora who studies traditional knowledge systems at the State University of New York College of Environmental Science and Forestry. "You know, most Westerners would never say that they felt Christianity influenced their science, or their views on wildlife, or their views on the world," he says. "Scientists, they would say it is totally divorced from that. And I don't think that's true. And I think it's just there in every single thing."

It's hard to see the water you swim in, the air you breathe. The culture you've grown up with. If you start assuming that humans are in charge, everything follows. If we're in charge, we have the power to destroy—and we also have the responsibility and the right to fix whatever we have changed, in whatever way we have to. Our blindness to our own cultural biases also makes it easy to pick apart other people's views of their ecosystems. Western scientists, Patterson says, "look at Indigenous traditional ecological knowledge and they go, 'but it's all wrapped up in their cosmology, it's all their religion.'" But those same scientists fail to see their own worldview is just as shaped by their beliefs. Yes, Patterson says, Indigenous knowledge is wrapped up in cosmology, "but at least they're honest about it."

Dominion, the idea that humans are destined to be here and

to be in charge, runs deep in our culture. Ian McHarg, a famous conservationist, once declared that all environmental degradation can be laid at the feet of the Bible. Genesis 1:28 lays it right out: "And God blessed them, and God said unto them, Be fruitful, and multiply, and replenish the earth, and subdue it: and have dominion over the fish of the sea, and over the fowl of the air, and over every living thing that moveth upon the earth." It is an idea that permeates most of Western culture—the idea that we, humans, have dominion over the environments around us, to do with as we see fit.

When people seek to do what they want with their surroundings, "I think there is a tendency to appeal to Genesis, and to that verse in particular," says Ken Stone, who studies Bible culture and hermeneutics (the fancy term for biblical interpretation) at the Chicago Theological Seminary. Because people reference it so often, he says, the idea of dominion has probably had an outsize influence on how we think about our relationship to the environment.

But the Bible does not stop at Genesis. "I think that [conservationists] are wrong to believe there's just one biblical view about the relationship between humans and animals," Stone says. "And I think having made that decision, they're wrong to emphasize the one that focuses on dominion and rule as the primary biblical view. It's a biblical view, yes, but it's not the only one."

Like those who turn to Genesis, most ethicists, if you really pressed them, would probably also put people above animals. If you give them the choice between saving a kitten and a human baby from a burning building, even the most staunch utilitarians, the ones who believe every species suffers equally, usually crack and pick the baby, says Nicholas G. Evans, a philosopher (and friend) at the University of Massachusetts in Lowell.

Ethics is not a personal system of right and wrong you feel in your gut. That's your moral center—the nougat of the soul. Instead, "most ethical systems are based on principles that you kind of figure out in advance," Evans explains. "And then you go out into the world and apply [them] and decide if things are good or bad."

Each ethical system has slightly different ways of weighing pros and cons. Broadly they fall into four buckets, says Evans. Some are based on consequences. "Find out what the good thing is in life and then make more of it," he says. Then there are duty-based systems, where actions are good or bad for their own sake. "So, murder bad, telling the truth good," Evans clarifies. "Even if telling the truth makes everyone much, much, much less happy, which sometimes it does."

The third bucket is about what kind of person you are. You can rescue all the kittens or babies you want—but if you do it for the wrong reasons? You're still not a good person. Finally, there's the fourth bucket which is, well, everything else.

All the different systems would probably agree that it's bad, say, to set a box of kittens on fire for no reason, Evans says. What they'll disagree on is the reasons why it's bad. Similarly, he says, most ethicists might well agree on a definition of the word "pest." A pest's presence is negative, and it would be a positive if the pest were gone. Ethicists would disagree, however, on what counted as a pest. Some would count rabbits, others wouldn't. Some probably claim no living creature could ever live up to the name, others might see pests everywhere they look.

And that matters, because once the label "pest" is in, thinking about its well-being is out. "You're telling the population that this animal is not worthy of as much consideration," says Francisco Santiago-Ávila, the conservation manager at Project Coyote.

If you can make the argument that this animal is harming you—better yet, that it's an existential danger to the survival of you or something you hold dear, from an endangered ecosystem to your garden, dog, cat, or cow—you can feel absolved of what you might have to do to get rid of the threat. It's so much easier to just call something a pest and make it killable. "Typically, we don't have a history of a systematic way of working through those ethical considerations," when it comes to the animals we hate, says Liv Baker, an animal behaviorist at Hunter College in New York City. "It's uncomfortable for one thing, and it forces us to, you know, potentially do things we don't want to do." If we had to look at pests as beautiful, sweet, or inspiring—like we look at, say, elephants, house cats, or horses—it suddenly becomes a lot harder to think they need to go. Similarly, when we see animals we care about and refuse to call them pests, it's a lot easier to excuse any damage they might cause. Especially when that damage is happening to people and ecosystems and things far away or out of sight.

Conservationists have warned us for a long time that we are predators on the landscape, the deadliest predators of all time. But when we confront a pest, it's all too easy to feel like prey. Even when the pest is a rat, or a mouse, or a pigeon in your face. They might be (in the case of mice) three thousand times smaller, but feel suddenly far too large. Emma and Paul were facing down a raccoon—an animal that averages between fourteen and twenty-three pounds. But they didn't see the size of a small dog. They saw teeth and claws and feared the pain of a bite.

To be prey, to be the constant victim of the nature you're trying to get away from, makes us anxious and twitchy. It makes us eager to defend, to take preemptive strikes. We, our crops, our animals, even other endangered species we care about, are all prey. Any threats need to be eliminated immediately.

Not all people, though, see themselves as prey. The Lakota don't. Instead, they are predators, along with the wolves, mountain lions, and bears. The first peoples, writes Joseph Marshall III in *On Behalf of the Wolf and the First Peoples*, emulated the hunting styles of wolves and other predators they interacted with. "But to be like the wolf meant that *they* also had to exist to serve the environment—to accept the mutuality of life." Every animal had their traits that helped them get by in the world, he notes. Deer have a strong sense of smell, and skunks have a strong odor. Porcupines have quills. People have reasoning, understanding. "In other words, the first peoples did not see their ability to reason or understand as anything that made them superior; instead, it was simply *their* key to survival."

This perspective is about living within an ecosystem. A view like this changes how you view other animals in "your" space, because the space isn't really yours alone. It's shared. "Our notions of treaty and relationship buildings . . . is first with the nonhumans," explains Darren Ranco, a citizen of the Penobscot Nation who is an anthropologist and Native American studies professor at the University of Maine in Orono. Animals are not simply things to be used that matter less—things that can be gotten rid of when inconvenient. Instead, Ranco says, people have to maintain and build treaties and relationships with the organisms around them.

These are not written treaties. "It's not like assigning them a document that is dead and kept somewhere," Ranco says. "The vibrancy of the relationship is very much tied to an ongoing set of relations and connections." Everything in the ecosystem—animals, plants, water, and more—has its own knowledge and its own responsibilities to fulfill. Nonhumans are treated as neighbors— beings that humans need to work with to get by in the world. If you fail to mow your lawn, let your garbage pile up, or routinely throw loud parties at 5:00 a.m., you're not being a good neighbor. Sometimes, the local homeowners' association, condo association, or police will let you know just how you are failing to keep your end of a good neighbor bargain.

This includes invasive species, Ranco notes. "Obviously, invasives are a problem for Indigenous people," he says, but in his research he has found that dealing with them isn't always about poisoning and killing them all. "It was this preference to maintain and build upon the broad interests of the ecosystem and not just respond as a kind of eradication approach," he explains. New species, or animals that are causing problems for people, are not the enemy. They are new neighbors—that need to be approached with new treaties. In an atmosphere of mutual responsibility, it's not an ongoing war between the harried human and the pernicious pest. It's about everyone involved collectively acting to keep the ecosystem running.

Even the word "pest" is one that some Indigenous groups just won't use. "We've been told specifically not to use the word 'pest' by tribal leadership," says Travis Bartnick, a wildlife biologist. Bartnick is not Indigenous but works for the Great Lakes Indian Fish and Wildlife Commission, which represents eleven Ojibwe tribes and their treaty rights for hunting, fishing, and gathering across Wisconsin, Minnesota, and Michigan. For the Ojibwe, he says, "all beings on this earth, they were put here by the Creator for a reason. And it's not their fault that they end up in these places. It's humans' fault that they end up in these places where they don't necessarily belong."

IN ROBERT SULLIVAN'S book *Rats*, he describes rats as "our mirror species, reversed but similar, thriving or suffering in the very cities where we do the same." Rats—pests in general—do, I think, hold up a mirror to our own lives. Pests often bring into relief what we hate about ourselves—our greed, our stubbornness, our fear, and the way we treat the things we disdain. They make us feel futile and powerless, yes. But they also show us the consequences of our own actions.

At this point, the obvious conclusion seems to be that humans—especially colonizing cultures with Western ideas of dominion over the landscape—are the real pest. We are the ones gathering in cities, invading habitat, providing garbage. We are the ones who brought voracious cats, rats, and more to vulnerable islands full of birds that can't fight back. We are the ones who toss garbage out and then look horrified when an animal treasures it. We are the ones who built a Panera and dared any animal to enter.

This answer, however, is too easy. It's the type of statement that makes people throw their hands up in defeat. What can we do? We are a scourge upon the planet!

Luckily for us, we are scrappy, resourceful, and able to change. We are not stuck with this war we have started. We can learn from each other—from other cultures and people and ways of living. "We need to acknowledge our role as pests and think about how are we contributing to this problem," says Anne Quain, the veterinary ethicist at the University of Sydney. "How are we creating the conditions for this population of animals to explode and addressing those conditions, rather than simply going after the animals?"

I began writing this book to find out why we call some animals pests, and some not. I wanted to find a new definition of pest. My original idea was to show that pests are a result of human perspective. A pest is only a pest because it bothers us, because it causes us harm. Pests aren't about the animals themselves, they're about us. I still do believe that. As I researched this book, however, poring over papers and books about rats and elephants and pigeons and toads and geese and monkeys and more, I began to realize something else.

I want to throw "pest"—and yes, it's the title of this book—away.

"Pest" makes our lives easy. But it doesn't serve us well. "Anytime we look at complex issues in black-and-white terms, I think it automatically puts us at a disadvantage," notes Kristina Slagle, an environmental social scientist at Ohio State University in Columbus. Pest puts an animal in a box. This animal is bad. It has no upside.

The word "pest" doesn't give animals—or us—a chance. It takes away context and complexity. It shuts down our curiosity, our desire to look for other angles or solutions. It puts animals in a box, yes. But it boxes in our thinking too.

So much of our surprise, our fear, and our vindictiveness when faced with a pest is the result of our own ignorance. When we see a coyote in the street, a rat in our trash can, or a squirrel in the attic, we are at a loss. When we don't know what to do, we feel helpless. Vulnerable. When we realize how helpless we are, shame follows immediately behind. We want the problem—and the animal causing our shame—to go away.

We could escape this mental mousetrap. We could learn about the animals living in our environments. We could learn the difference between a mouse, a vole, and a rat. We could learn about the ecology of the place we live, and understand how we impact our environments and how those environments adapt in return.

Some of this knowledge will come from people like the many wildlife scientists who generously taught me as I wrote this book. They spend their lives in pursuit of deeper knowledge of the animals around us. Some will come from people like the sociologists and psychologists whom I spoke to, who work to understand why people think the way they do, and how best to change our behavior and make it stick.

And some will come from Indigenous people all over the world who have spent thousands of years living with and observing the animals around them—from wolves to elephants and beyond. Several have been generous and patient in sharing that knowledge with me. This is knowledge that Indigenous people have chosen to learn and cultivate, chosen to keep alive.

This is knowledge that many of us who grew up steeped in Western culture walled ourselves off from—and even prevented Indigenous people from preserving. Do we even need to know about raccoons, squirrels, or deer? What about rats or pigeons?

If we want to live with them—and not against them—then I am convinced that the answer is yes. If we want to enjoy wildlife near us, we need to learn that it won't stay where we put it. And we need to know where it's likely to go, what it's likely to need, and why it might do what it does.

This doesn't mean we all need to take to the woods and forage for nuts and berries. It doesn't mean we need to all become hunters or welcome birds into our homes. Instead, it means we need to listen to Indigenous people, and take their knowledge and viewpoints seriously. We need to learn from scientists and from our own surroundings about the ecology of where we live.

"I think that our job is to revive these instructions across all human beings," Patterson says. The Haudenosaunee (a confederacy containing the Tuscarora) have a prophecy that "someday people will come to us, and they'll say, what are your prophecies? And how do we transform our way of thinking? And that seems to be happening, I could be wrong, but it's certainly happening more now than when my parents grew up." People are realizing that the way we view ourselves, and our relationship to where we live, isn't the only way—or the best one. It's more than time to pay attention to the knowledge of groups that have been ignored and marginalized.

Patterson doesn't think about himself, the Tuscarora, or other Indigenous people as having specific Indigenous knowledge. He prefers to call it naturalized knowledge, knowledge that can be shared, so everyone can change. "Because these are instructions for everyone."

NONE OF THE ideas in this book are new ones, especially when we take into account Indigenous views; ideas of coexisting with wildlife are thousands of years old. Western science has caught up too. Wildlife managers have for a long time now talked about turning conflict into coexistence.

The word "coexistence" conjures up pretty images of people having a picnic under a blue sky in a park while raccoons and deer romp peacefully nearby. The lion lies down with the lamb, for good measure. More important, the people are doing exactly what they like, and the animals are behaving exactly as we think they should.

If that's what we think coexistence is, we will never, ever get there.

When I think of coexistence, I think of walking the streets of a city, driving, going grocery shopping. There are so many things that you do without thinking—things that allow you to live in a society with other people. You walk on the sidewalk instead of the street. That's for safety, but it's also because if we want this society to work, people need to get where they're going without interference—whether walking, biking, or driving. We behave in predictable ways—stopping at stoplights, moving to one side so someone can pass. We wait in line at the store. In exceptionally busy cities, we even give people the courtesy of not seeing them—of walking right past someone walking the other way, giving someone privacy in public. We do this because we don't exist in conflict with other humans (most of the time). We coexist with them.

There's no one silver bullet here. No one thing that will make all pests, or even a single pest, leave us alone. Instead, what's needed is a network of changes that acknowledge that animals live with us, instead of viewing them as constant interlopers on our world. Society is, at its basis, a set of rules that we all obey that allow for mutual coexistence. If we added animals—some of which we think of as pests—to what we expected to deal with every day, what would it look like? It would look like some different rules, some different ways of behaving. We lock doors against other humans; we get better trash cans (and use them), and "lock our doors" against mice, rats, bears, and raccoons, stopping them getting in. We could provide more people with solid, clean, comfortable places to live. We could give other animals a wider berth and drive slower, instead of expecting them to recognize a car going seventy miles an hour. We could keep our dogs on leashes. We could keep our cats inside. We could watch our feet, because goose poo happens.

We could acknowledge the habitat we create for deer. We could either bring their other predators back—or add venison to our diets. Pigeons and rabbits linger on in a niche we once appreciated, as food. We could appreciate them again.

One of the things that allows society to go on in the way it does is the consistency of our behaviors. When people don't stop at a red light, that is bad. No one supports that. If people cut the line, we've got something to say. Similarly, coexistence with urban wildlife would need consistency. Right now, any animal encountering people from a Westernized society faces a potentially deadly roulette. Some will feed them and fawn on them. Some will scream and run the other way, and some will shoot or trap. If we truly want to coexist and not just live near wildlife? Our reactions can't be individualistic and about making personal connections. We might have to do the unthinkable—we might have to leave wildlife unfed.

We also need knowledge. We put rules into place to make society run, knowing that the majority of people are capable of learning and following them. To coexist with elephants, people in Kenya have learned both over time and with scientific methods a vast array of nonlethal techniques. They employ those techniques—hazing and honeybees and helicopters—as consistently as they can. To coexist with rats and pigeons, we will need to learn about them. We will need to study bears, turkeys, pythons. To coexist, we need a deep understanding of what we're working with.

The final thing we need is humility. We will need to acknowledge that existing with animals cannot be done without some inconvenience—and even difficulty and pain. Coexistence isn't always peaceful and sweet. In Western countries we are very used to a painless natural experience, where any loss—of a pet, a crop, or a chicken—is unacceptable. But coexistence means that some loss will happen, as pig farmer Robin Sage said. Even with his giant dogs and electric fencing, he pays in, sometimes, to the Red Bird Acres owl fund.

People coexist with elephants, but they know that it is not always painless or easy. Sometimes it is very hard. But the people who live with elephants believe they are worth coexisting with—and their responses reflect that. Coexisting in a society with pests would mean giving up some of our power, acknowledging there are some things we cannot always control. Our flocks and harvests will require protection—guard dogs, scaring lights, or flags. We and our pets won't roam in the woods without a care.

IT IS PERHAPS fitting that, in the final days of writing this book, I got attacked by a wild turkey.

I was out for a run by the Anacostia River on a leaden, chilly day in December. I admit I was totally zoned out as I rounded a bend and confronted three turkeys crossing the path. The first two were females, the third a male, judging by the blue-and-red wattle that graced his face. The females had mostly crossed the path, and the male paced protectively behind. I slowed a little to give him time, and then ran by on the other side of the path, fairly sure I had given him a wide enough berth.

I hadn't. The turkey turned swiftly and issued a gobble that was clearly intended to be menacing (sadly, no gobbling sound is ever really menacing). He began to stalk after me as I ran by.

When I glanced again, the turkey had transformed. When crossing the road he had been small and sleek, only about as high as my knee, the wattles the only giveaway that this turkey was a tom. Now, though, his blood was up. He stood up, his tail fanned out and his wings fluffed and he was suddenly twice the size, rising almost to my hip. This was a turkey reminding me that he had dinosaurs in the family. And he kept coming, running after me.

I yelled a loud "No! Git!" and ran faster. I figured if I got out of the range of his ladies he'd settle down.

He did not settle down. He ran faster too, racing after me and gobbling. He started making threatening rushes, flapping and back-winging to reach forward at me with long, wicked claws. Turkey toes are no joke. But I realized abruptly that this wild Butterball was hazing me. He didn't ever get too close—carefully remaining out of my reach. He was staying low. He was trying to see me off.

I tried to leave. But no matter how I yelled, assuring him I was only attracted to turkey breasts in sandwich form, and no matter how fast I ran, this turkey was not giving up. After about a quarter mile of this, I spotted what I needed. As he lunged at me again, I darted to the side of the path and hefted a large stick. I stopped and faced down Thanksgiving dinner.

Who's Big Bird now? The turkey's whole demeanor changed. Even the look in his eye, I would swear, went from red rage to wide-eyed dismay. He had clearly forgotten that animals like me might not have claws—but we do have opposable thumbs. And I hazed that bird right back. I yelled and waved my stick, and he backed away. He tried one more rush, and I lunged in return. His plumage went comically flat, and he slunk back into the woods. I walked backward for a bit with the stick until I was fully out of sight, and then continued on, warning other runners that a turkey down the road was having some big feelings. Maybe carry a stick.

The absurdity of the whole thing hit me immediately. We must have looked hilarious. But when I ended up writing a story about the bird and his many attacks on runners, bikers, and walkers, I found that many people responded with panic. Wild turkeys were foreign to their urban experience.

They won't be for long. Indigenous people were raising turkeys (*Meleagris gallopavo*) in what is now Mexico possibly as early as 300 BCE. Of course, when Europeans arrived, they immediately hunted them and hacked down their habitat. The birds were totally gone from New England by the end of the Civil War. As Jim Sterba recounts in *Nature Wars*, after reintroduction in the 1950s, flocks in the United States took off. Like white-tailed deer, they thrive in the combination that humans are really good at building—woods and open lands, nicely patchworked together. By 2000 there were more than four million birds.

Then the complaints began. Turkeys love seeds and seedlings and farmers complained of stolen wine grapes and hay ruined by scratching turkey claws. The birds were also bold. When I first saw wild turkeys in a city—parading grandly down Massachusetts Avenue in Cambridge—my friends and I were thrilled. Soon, though, we started hearing how turkeys would chase children and pets. In 2020, a large flock (also called a rafter) took up residence at the NASA Ames Research Center in Moffett Field, California, and started reminding NASA that rocket scientists cannot, in fact, solve everything. Suddenly, turkeys were a little too much nature, a little too close. They were unexpected—and more importantly, people didn't know what to do. They might run, they might look big and haze the turkeys. They might leave out food. People reacted to the turkeys with all the knowledge they had about the local wildlife, which was—often—almost none at all.

Our reactions to the animals in our lives are often a wild seesaw of deadly conflict and cooing compassion. Some people feed deer and want to pet them. Others desperately chase them out of their yards to protect their dogs. Some people dump trash for rats to rustle in, and then those same people might lay out poison to kill them. Some people feed pigeons, others put out spikes to keep them away. Some people love bird feeders, and others come and trap the bears that spend too much time tearing down bird feeders and breaking into garages full of second refrigerators and birdseed.

For a brief, tiny second, there might be balance as the seesaw swings—a moment where everyone's feet are on the ground. Where we see animals around us, acknowledge them, and change our behavior. But most of the time, the seesaw is swinging up and down. The essence of human-wildlife conflict is that there is so little balance, so little tolerance that acknowledges that nature cannot exist without in some way affecting the way we live our lives.

For every person who hazes the turkey, another might spread corn to feed it and take video of the gorgeous fowl. As it happened, that particular turkey became a menace, haunting the running trail and producing plenty of videos as he attacked bikers, runners, and walkers. The local wildlife biologists and park rangers put on a hunt for the turkey to relocate it. They never managed to catch him, and he's apparently disappeared for now. But if he comes back, if his temper tantrums continue, people will quickly lose their patience. The turkey will end up with its legs as Renaissance Faire food, another ex-pest.

I hazed that turkey, and I tried to ensure that other people would too. It wasn't just for me (though it definitely was for me). It was for him. Both he and we need to make sure no one gets too comfortable. We need to keep our guard up, and know the rules, to know what a turkey's behaviors mean. We need to keep our reactions to him predictable, to make his behaviors predictable in return. We need to see these animals as part of our society, sharing our red lights and our sidewalks. I took a lesson in humility from him—turkeys can mean business. I'll be giving him plenty of space in the future. He might change from our encounter. But if we're going to live together, I will need to change too.

Nature isn't always going to be tame and neutered for our pleasure. It won't be out there, where we need to drive to it. It runs through our walls and in our sewers. It flutters over our streets and picks its way through our yards. It eats our trash and our crops. It takes advantage of our weakness. We need to learn there's more than one way to be strong.

And as I've shown throughout this book, we can't beat nature, not really. Our efforts to come out on top may work for a house, or even an island. But humans are everywhere, and where we are, "pests" are too. Our behaviors and possessions create our own eco-system, a place where animals can live off us.

If we want nature in our lives, want to live near the wild, we're going to have to change how we look at ourselves. Only by under-standing the animals around us, by learning what attracts them and what repels them, can we live with them, instead of against them. We cannot dominate and squash the world into the boxes of our preferences and definitions. But we can accept it, we can adapt. We can protect ourselves with knowledge—and stop going out looking for a fight.

ACKNOWLEDGMENTS

This book was written in large part on the unceded lands of the Nacotchtank (Anacostan) people, with other chunks written on the unceded lands of the Kalapuya and the Nonotuck. If I've learned anything writing this book, it's how crucial Indigenous knowledge holders everywhere are to changing our interactions with our environments. It is up to non-Indigenous people to listen, learn, and put in the work.

The best part of the book-writing process is the opportunity to indulge in every single impulse of my never-ending curiosity, and I am so grateful for everyone who enables me. First, my agent, Alice Martell, is the best hypewoman a writer could have. My editors, Gabriella Doob, Norma Barksdale, Jane Cavolina, and Denise Oswald, thank you for putting your faith in this project and for saving me from myself. My fact-checker, Kyle Plantz, thank you for your eagle eyes and well-honed skills. Any mistakes remaining are my own fault.

Big projects like this require space, time, and skill. For the space and time, the Knight Science Journalism program at MIT under the direction of Deborah Blum and Ashley Smart was absolutely essential. Not only did it help me take courses to gain expertise and give me the time to write the proposal for this book, it introduced me to the 2019–2020 KSJ Class. Anil Ananthaswamy, John Fauber,

Andrada Fiscutean, Richard Fisher, Tony Leys, Thiago Medaglia, Sonali Prasad, Molly Segal, and Eva Wolfangel are a group like no other, and helped me so much as my ideas came together.

For the skill, I turn to the team at *Science News* and *Science News Explores*. Nancy Shute, Sarah Zielinski, Kate Travis, and the entire team are outstanding colleagues and taught me everything about honesty, integrity, and straight-shooting news (but will never convince me to give up the Oxford comma). Janet Raloff in particular gave me a chance, and has made me the journalist I am today. The lovely podcasting team at Science for the People helped hone my interview skills.

Many wonderful sources and experts took the time to educate me and to share their expertise. Some gave an hour, some let me into their classes, and still others educated me in ways great and small. There are nearly three hundred of them. Many did not end up in the pages of this book, but their ideas all helped to form its words. (Stretches neck, takes deep breath.) In alphabetical order they are: Sunandan Adha, Shelley Alexander, Alexis, Thom Almendinger, Rachel Ankeny, Fabian Aubret, Amy Bachman, Carrie Baker, Liv Baker, Katherine Barnhill-Dilling, Heather Barr, Travis Bartnick, Jon Beckmann, Marc Bekoff, Beth Berkowitz, Carrie Bickwit, Dawn Biehler, Mark Biel, Michael Blum, Ben Bolker, Brad Bolman, Jason Boulanger, Jonathan Boyar, Gustavo Bravo, Mary Brazleton, Stewart Breck, Garrett Broad, Justin Brown, Kristin Brunk, Jeremy Bruskotter, Henry Buller, Kaylee Byers, Karl Campbell, Elizabeth Carlen, Colin Carlson, Carrie, Sarah Carroll, Miguel Chevere, Tom Chiller, Sonja Christensen, Chris Ciuro, Pam Comeleo, Randy Comeleo, Kim Cooper, Jayna Corns, Bobby Corrigan, Michael Cove, Phillip Cox, Sarah Crowley, Thomas Cucci, Joachim Dagg, Peter David, Katharine Dean, Mícheál De Barra, Kristen Denninger Snyder, Chris DePerno, Tristan Donovan, David Drake, Don Driscoll, Alex Dutcher, Bahar Dutt, Scott Edwards, Susan Elias, Emma, Kevin Esvelt, Nick Evans, Fa-Ti Fan, Rowan Flad,

Richard Forman, Jane Foster, Camilla Fox, Jitudan Gadhavi, George Gallagher, Rafael Garcia, Stan Gehrt, Madeleine Geiger, Tom Gilbert, Jacquelyn Gill, David Givens, Andreas Glanznig, Jenny A. Glikman, John Godwin, Eugene Goldfarb, Meredith Gore, Ashley Gramza, Daniel Grear, Miriam Gross, Lori Gruen, Anja Guenther, Anita Guerrini, Fatima Guled, Catherine Hall, Samneiqua Halsey, Rebecca Hardesty, Kristen Hart, Dora Henricksen, Steve Henry, Hal Herzog, Joe Hinnebusch, Hopi Hoekstra, Matthew Holmes, Alice Hovorka, Ardern Hulme-Beaman, Melissa Jenkins, Colin Jerolmack, Heather Johnson, Danson Kaelo, Donna Kalil, Elinor Karlsson, Martin Kavaliers, Roland Kays, Chris Kelty, Bruce Kimball, Barbara J. King, Lucy King, Fabienne Krauer, Suresh Kuchipudi, Carl Lackey, Max Lambert, Felix Landry Yuan, Kelly Lane-DeGraaf, Crystal Lantz, Greger Larson, Louis Lefebvre, Sarah Legge, Kirsten Leong, Julie Levy, Matt Liebmann, Anna Lindholm, Wayne Linklater, Lauren Lipsey, Stacy Lischka, Anastasia Livtinsteva, Vanessa LoBue, Mark Long, Kathryn Lord, William Lynn, Christos Lynteris, Suzanne MacDonald, Peter Marra, Lynn Martin, Danielle Martinez, bethany ojalehto mays, Richard Meadow, Raul Medina, Guy Merchant, Alan Mikhail, Anders Møller, Javier Monzon, Lisa Moses, Scott Mullaney, Asia Murphy, Maureen Murray (Boston), Maureen Murray (St. Louis), Melani Nardone, The National Library of Aotearoa/New Zealand, Lisa Naughton-Treves, Victor Ndombi, Nicole Nelson, Chase A. Niesner, Camilla Nord, Phillip Nyhus, Colleen Olfenbuttel, Eileen O'Rourke, David Orton, Kriston Pape, Tom Parr, Michael Parsons, Paul, Jackson Perry, Anna Peterson, Vanessa Petro, Jared Piazza, Anna Pidgeon, Ray Pierotti, James Pokines, Karni Pratap, Kate Pritchett-Corning, Laura Prugh, Emily Puckett, Anne Quain, Niamh Quinn, Maud Quinzin, Karen Rader, Thomas Rawinski, Jennifer Raynor, Paul Rego, Jonathan Richardson, Jeurgen Richt, Harriet Ritvo, Andrew Robichaud, Joshua Rottman, Paul Rozin, James Russell, Allen Rutberg, Royden Saah, Wilson Sairowua, Francisco Santiago-Ávila, Julie Savidge,

Chris Schell, Boris Schmid, Manon Schweinfurth, Esther Serem, James Serpall, Andy Sheppard, Rick Shine, John Shivik, Anne Short Gianotti, Susan Shriner, Shane Siers, Georgiana Silviera, Palatty Sinu, Kristina Slagle, Kirsty Smith, Will Smith, Carly Sponarski, Felicia Staley, Ted Stankowich, Ken Stone, Daniel Storm, Tanja Strive, Jennifer Strules, Jacqueline Sullivan, Mingli Sun, Marliese Talay, Sam Telford, Lydia Tiller, Sarra Tlili, Adrian Treves, Ashley Valm, Mark Vieira, Kathy Vo, Susanne Vogel, Bridgett vonHoldt, Jake Wall, Arian Wallach, Feiran Wang, Derick Wanjala, Jennifer Ward, Georgia Ward-Fear, David Watson, Maggie Watson, Dave Wattles, Matthew Webster, Sam Weiss Evans, Lior Weissbrod, Elic Weitzel, Margaret Wild, Adam Wilkins, Matti Wilks, Fang Xiaoping, Julie Young, Malinda Zeder, Kathy Zeller, and Zhibin Zhang.

I would particularly like to acknowledge the Indigenous peoples across several countries who spoke with me, educated me, or provided resources. I so appreciate the time, patience, and teachings of Nasbah Ben, Karen Bennally, Bradford Haami, Samuel Kala, Martin Maina, Joseph Marshall III, Danica Miller, Douglass Neasloss, Jonah Noosaron, Hori Parata, Neil Patterson, Darren Ranco, Mere Roberts, Eli Suzukovich, Purity Taek, and Harry Walters (who led me to the work of Steve Pavlik).

A researcher is only as good as her translators. Thank you so much to Yenting Chen, Dinah Gardner, Haze Liff, and Simon Mwanza for lending their language skills. Thanks also to my sensitivity readers: Alicia Gangone, Rim Kazhall, Joseph Lee, Simon Mwanza, Sonali Prasad, Giselle Routhier, and Zhutian Yang. And my wonderful expert readers: Rodrigo Pérez Ortega, Jonathan Richardson, Molly Segal, and Susanne Vogel. Your honesty has made both me and the book better.

Colleagues have also helped to soothe my near constant anxiety and encouraged me to embrace the sheer joy of reporting. Maryn McKenna, Christie Aschwanden, Emily Willingham, Shannon

Palus, Amy Maxmen, Adam Rogers, Brendan Maher, Rebecca Boyle, Betsy Mason, Jeffrey Perkel, Tim De Chant, Zach Zorich, Annalee Newitz, David Bamford, and Tina Saey, thank you so much for your support and friendship.

Many scientists and journalists can end up being defined as people by their profession. It is thanks to my wonderful families, by birth and choice, that I am not. To the Capitol Hill Chorale, DC Judo, and the Bruces, thank you for reminding me who I am. Jyoti Daniere keeps me (relatively) sane. Tim Fothergill and K. O. Myers, Sarah Brett England, and Ericka MacLeod gave constant support and kept the alcohol supplies high. Doug McNamara motivated me with coffee and David Steussey with long runs. Every day, Kendra Pierre-Louis, Mika McKinnon, and Kelly Hills have been there, my friends in my pocket and my heart. Riley Black believed in me as a writer before I ever believed in myself. Shannon Griswold, I'd travel with you to the ends of the earth.

My brother, David, keeps my debate skills honed. My mother Cathy's boundless faith keeps me going. My father, Bob, catches my every tpyo (that one's for you, Dad). My partner, Vince, none of this would be possible without you. I'm the luckiest human on earth, and H. H. Boots, Purrveyor of Dry Goods, and Eliza Schuyler Hamilton are the luckiest cats.

Finally, to F**king Kevin: You inspire me, dude. I was kidding about the mange. I'll leave a tomato out in your honor.

NOTES

INTRODUCTION: A PEST IS _____?

xii   lean times in winter: A. Brodin, "The History of Scatter Hoarding Studies," *Philosophical Transactions of the Royal Society of London B: Biological Sciences* 365 (2010), doi: 10.1098/rstb.2009.0217.

xii   months after hiding them: I. M. V. MacDonald, "Field Experiments on Duration and Precision of Grey and Red Squirrel Spatial Memory," *Animal Behaviour* 54 (1997), https://doi.org/10.1006/anbe.1996.0528.

xiii disheveled deer parks: M. Holmes, "The Perfect Pest: Natural History and the Red Squirrel in Nineteenth-Century Scotland (William T. Stearn Prize 2014)," *Archives of Natural History* 2, no. 1 (2015), https://doi.org/10.3366/anh.2015.0284.

xiii Victorian egg collector: Holmes, "The Perfect Pest."

xiii bird-egg munching villains: Holmes, "The Perfect Pest."

xiii and the English: P. Coates, "A Tale of Two Squirrels: A British Case Study of the Sociocultural Dimensions of Debates over Invasive Species," in *Invasive Species in a Globalized World: Ecological, Social, and Legal Perspectives on Policy*, ed. Reuben P. Keller, Marc W. Cadotte, and Glenn Sandiford, (Chicago: University of Chicago Press, 2014), doi: 10.7208/chicago/9780226166216.001.0001.

xiv   was introduced from the Americas: Coates, "A Tale of Two Squirrels."

xiv   cause for conservation concern: Holmes, "The Perfect Pest."

xv   "Human-Wildlife Conflict and Coexistence": Nyhus, "Human-Wildlife Conflict and Coexistence."

xvi   Cats kill between: S. Loss, T. Will, and P. Marra, "The Impact of Free-Ranging Domestic Cats on Wildlife of the United States," *Nature Communications* 4 (2013), https://doi.org/10.1038/ncomms2380.

xvi   eight quadrants on a graph: P. J. Nyhus, "Human-Wildlife Conflict and Coexistence," *Annual Review of Environment and Resources* 41 (2016): 143–71, https://doi: https://doi.org/10.1146/annurev-environ-110615-085634.

xvi cubes of human judgment: Nyhus, "Human-Wildlife Conflict and
     Coexistence."
xx   like tiny mailboxes: E. Cortesi, et al., "Cultural Relationships Beyond
     the Iranian Plateau: the Helmand Civilization, Baluchistan and the
     Indus Valley in the 3rd Millennium BCE," *Paléorient* 34 (2008), doi:
     10.2307/41496521.
xx   "images of your tumors": 1 Samuel 6:5.
xxi deserved the same fate: "The Farmer and the Stork," *Aesop Fables*.
xxii fear and loathing: No matter where they are found, gray wolves the world over
     are *Canis lupus*, and can and do interbreed with domestic dogs. Nature isn't far
     away after all.
xxii run out of wolves: J. Strutt, *The Sports and Pastimes of the People of England
     from the Earliest Period, Including the Rural and Domestic Recreations, May
     Games, Mummeries, Pageants, Processions and Pompous Spectacles* (London:
     Methuen, 1903).
xxii another two centuries: Strutt, *The Sports and Pastimes of the People of England*.
xxiii with a stick in 1621: J. T. Coleman, *Vicious: Wolves and Men in America* (New
     Haven, CT: Yale University Press, 2006).
xxiii for their own meals: "Wolf Wars: America's Campaign to Eradicate the Wolf,"
     The Wolf That Changed America, season 24, episode 4, PBS, September 14,
     2008.
xxiii named the world: "Ma'iingan (The Wolf) Our Brother," White Earth Land
     Recovery Project, https://www.welrp.org/about-welrp/maiingan-the-wolf-our
     -brother/.
xxiv "join them on earth": "Ma'iingan (The Wolf) Our Brother."
xxiv the lower forty-eight states: "Wolf Wars."
xxiv wolves, grizzly bears: J. A. Estes, et al., "Trophic Downgrading of Planet Earth,"
     Science 333 (2011), doi: 10.1126/science.1205106.
xxiv Colorado, Idaho, and Montana: S. Brasch, "It's Official: Colorado Has Its First
     Wild Wolf Pups Since the 1940s," CPR News, June 9, 2021.
xxvi Biehler is the author of: D. Biehler, *Pests in the City: Flies, Bedbugs,
     Cockroaches, and Rats* (Seattle: University of Washington Press, 2013).
xxvii lack of opportunity, or simple poverty: Biehler, Pests in the City.

CHAPTER 1: A PLAGUE OF RATS

4   known as a sage: "About Karni Mata," Karni Mata Temple, http://
     matakarnitemple.com/karni-mata/, accessed May 17, 2022.
7   Romans left Britain: K. Rielly, "The Black Rat," in *Extinctions and Invasions:
     A Social History of British Fauna*, ed. T. O'Connor and N. Sykes (Oxford, UK:
     Oxbow Books, 2010), 134–45.
8   new ideas about those rats: Jonathan Burt, *Rat* (London: Reaktion Books, 2005).
8   sewers began to proliferate: Burt, *Rat*.

8   recognized across cultures: P. Ekman and W. V. Friesen, "Constants Across
      Cultures in the Face and Emotion," *Journal of Personality and Social
      Psychology* 17 (February 1971), http://doi.org/10.1037/h0030377.

10   drug called domperidone: C. L. Nord, et al., "A Causal Role for Gastric
      Rhythm in Human Disgust Avoidance," *Current Biology* 31 (February 2021),
      https://doi.org/10.1016/j.cub.2020.10.087.

12   SGARs for rodent control: California State Assembly Bill No. 1788, Pesticides:
      Use of Second Generation Anticoagulant Rodenticides, September 30, 2020,
      https://leginfo.legislature.ca.gov/faces/billTextClient.xhtml?bill
      _id=201920200AB1788.

13   and 13 deaths: "Leptospirosis in Dogs in Los Angeles County in 2021," County
      of Los Angeles Public Health, Veterinary Health, updated March 11, 2022,
      http://www.publichealth.lacounty.gov/vet/Leptospirosis2021.htm.

13   third plague pandemic: J. Frith, "The History of Plague—Part 1: The
      Three Great Pandemics," *Journal of Military and Veterans' Health* 20
      (April 2012).

14   skeleton from Latvia: J. Susat, "A 5,000-Year-Old Hunter-Gatherer Already
      Plagued by *Yersinia pestis*," *Cell Reports* 35 (June 29, 2021), https://doi
      .org/10.1016/j.celrep.2021.109278.

14   skeleton from Russia: M. A. Spyrou, et al., "Analysis of 3800-year-old *Yersinia
      pestis* Genomes Suggests Bronze Age Origin for Bubonic Plague," *Nature
      Communications* 9 (June 2018), https://doi.org/10.1038/s41467-018
      -04550-9.

15   the main drivers: K. R. Dean, et al., "Human Ectoparasites and the Spread of
      Plague in Europe during the Second Pandemic," *Proceedings of the National
      Academy of Sciences* 116 (January 2018), https://doi.org/10.1073
      /pnas.1715640115.

16   Some are pneumonic: D. J. D., Earn, et al., "Acceleration of Plague Outbreaks
      in the Second Pandemic," *Proceedings of the National Academy of Sciences* 117
      (October 2020), https://doi.org/10.1073/pnas.2004904117.

16   for rat tails: Michael Vann and Liz Clarke, *The Great Hanoi Rat Hunt: Empire,
      Disease, and Modernity in French Colonial Vietnam* (New York: Oxford
      University Press, 2019).

17   had the villages destroyed: Maurits Bastiaan Meerwijk, "Bamboo Dwellers:
      Plague, Photography, and the House in Colonial Java," in *Plague Image and
      Imagination from Medieval to Modern Times, Medicine and Biomedical Sciences
      in Modern History*, ed. C. Lynteris (Palgrave Macmillan, Cham), https://doi
      .org/10.1007/978-3-030-72304-0_8.

19   Predator Free 2050 strategy: "Predator Free 2050," Department of
      Conservation, Te Papa Atawhai, Government of New Zealand, https://www.doc
      .govt.nz/nature/pests-and-threats/predator-free-2050/.

19   Māori to trap them again: P. M. Wehi, et al., "Managing for Cultural Harvest of
      a Valued Introduced Species, the Pacific Rat (*Rattus exulans*) in Aotearoa New

Zealand," *Pacific Conservation Biology* 27 (August 2021), https://doi .org/10.1071/PC20094.

20 Robert Sullivan's book: Robert Sullivan, *Rats: Observations on the History & Habit of the City's Most Unwanted Inhabitants* (New York: Bloomsbury, 2004).

22 a single bin cost: "A Look at UM's New $4,500 Trash Cans," *Daily Mississippian*, January 16, 2019, https://thedmonline.com/a-look-at-ums-new -4500-trash-cans/.

25 interviewed twenty people: K. A. Byers, et al., "'They're Always There': Resident Experiences of Living with Rats in a Disadvantaged Urban Neighbourhood," *BMC Public Health* 19 (July 2019), https://doi.org/10.1186 /s12889-019-7202-6.

27 did just fine: Michael H. Parsons, et al., "Rats and the COVID-19 Pandemic: Considering the Influence of Social Distancing on a Global Commensal Pest," *Journal of Urban Ecology* 7, no. 1 (September 2021), https://doi.org/10.1093 /jue/juab027.

28 like Times Square: "NYC Trash Bin Pilot Program Aims to Curb Large Garbage Piles on City Streets," Eyewitness News, ABC, April 20, 2022, https:// abc7ny.com/eric-adams-new-york-city-boroughs-waste-bins/11773111/.

CHAPTER 2: A SLITHER OF SNAKES

30 slithering through South Florida: "Burmese Python," Everglades National Park, National Park Service, updated August 12, 2021, https://www.nps.gov/ever /learn/nature/burmese-python.htm.

30 absence of rabbits: M. E. Dorcas, et al., "Severe Mammal Declines Coincide with Proliferation of Invasive Burmese Pythons in Everglades National Park," *Proceedings of the National Academy of Sciences* 109 (January 2012), https://doi .org/10.1073/pnas.1115226109.

33 participating in the Florida Python Challenge: "Florida Python Challenge 2022," https://flpythonchallenge.org/, accessed May 17, 2022.

33 removed 223 snakes: "223 Pythons Removed during 2021 Florida Python Challenge," Spectrum News, Bay News 9, August 4, 2021, https://www .baynews9.com/fl/tampa/news/2021/08/04/223-pythons-removed-during-2021 -florida-python-challenge.

34 have a healthy concern: G. C. Davey, "Self-Reported Fears to Common Indigenous Animals in an Adult Uk Population: The Role of Disgust Sensitivity," *British Journal of Psychology* 85 (November 1994), http://www.doi .org/10.1111/j.2044-8295.1994.tb02540.x.

35 babies will focus intently: V. LoBue and K. E. Adolph, "Fear in Infancy: Lessons from Snakes, Spiders, Heights, and Strangers," *Developmental Psychology* 55 (September 2019), http://doi.org/ 10.1037/dev0000675.

35 born to fear snakes: V. LoBue, et al., "Young Children's Interest in Live Animals," *British Journal of Developmental Psychology* 31 (March 2013), https:// doi.org/10.1111/j.2044-835X.2012.02078.x.

35   some side-eye: Susan Mineka, et al., "Fear of Snakes in Wild- and Laboratory-Reared Rhesus Monkeys (*Macaca mulatta*), *Animal Learning & Behavior* 8, no. 4 (1980): 653–63, https://doi.org/10.3758/BF03197783.

35   not at all: J. Joslin, H. Fletcher and J. Emlen, "A Comparison of the Responses to Snakes of Lab- and Wild-Reared Rhesus Monkeys," *Animal Behaviour* 12, nos. 2–3 (April–July 1964): 348–52, https://doi.org/10.1016/0003 -3472(64)90023-5.

35   pick up on their nerves: S. Mineka, et al., "Observational Conditioning of Snake Fear in Rhesus Monkeys," *Journal of Abnormal Psychology* 93, no. 4 (1984): 355–72, https://doi.org/10.1037/0021-843X.93.4.355.

36   pointier snake as mean: J. Souchet and F. Aubret, "Revisiting the Fear of Snakes in Children: The Role of Aposematic Signalling," *Scientific Reports* 6, November 2016, https://doi.org/10.1038/srep37619.

37   a fearful voice: LoBue and Adolph, "Fear in Infancy."

37   might kiss them: M. Conrad, L. B. Reider, and V. LoBue, "Exploring Parent–Child Conversations about Live Snakes and Spiders: Implications for the Development of Animal Fears," *Visitor Studies* 24 (February 2021), doi: 10.1080/10645578.2020.1865089.

37   provide negative information: Conrad, Reider, and LoBue, "Exploring Parent–Child Conversations."

38   god of the underworld in Europe: Emma Marris, *Wild Souls: Freedom and Flourishing in the Non-Human World* (New York: Bloomsbury, 2021).

38   snakes in northern Europe: H. J. R. Lenders and I. A. W. Janssen, "The Grass Snake and the Basilisk: From Pre-Christian Protective House God to the Antichrist," *Environment and History* 30 (August 2014), http://doi.org/10.3197 /096734014X14031694156367.

39   sixty of those: C. Arnold, "Snakebite Steals Millions of Years of Quality Life in India," *Nature News*, December 4, 2020, https://www.nature.com/articles /d41586-020-03327-9.

39   averaged 58,000 deaths: Z. E. Selvanayagam, et al., "ELISA for the Detection of Venoms from Four Medically Important Snakes of India," *Toxicon* 37 (May 1999), https://doi.org/10.1016/S0041-0101(98)00215-3.

40   101 snakebite deaths: S. C. Greene, et al., "Epidemiology of Fatal Snakebites in the United States 1989–2018," *American Journal of Emergency Medicine* 45 (July 2021), https://doi.org/10.1016/j.ajem.2020.08.083.

40   neglected tropical disease: "Snakebite Envenoming," World Health Organization, May 17, 2021, https://www.who.int/news-room/fact-sheets/detail /snakebite-envenoming.

41   real, live ones: F. L. Yuan, et al., "Sacred Groves and Serpent-Gods Moderate Human-Snake Relations," *People and Nature* 2 (March 2020), https://doi .org/10.1002/pan3.10059.

41   about 16 percent: Yuan, et al., "Sacred Groves and Serpent-Gods."

42   portions of the island: T. H. Fritts, et al., "Symptoms and Circumstances Associated with Bites by the Brown Tree Snake (Colubridae: *Boiga irregularis*)

on Guam," *Journal of Herpetology* 28 (March 1994), https://doi
.org/10.2307/1564676.

43 taking up oxygen: L. Clark, C. Clark, and S. Siers, "Brown Tree Snakes:
Methods and Approaches for Control," in *Ecology and Management of
Terrestrial Vertebrate Invasive Species in the United States*, ed. W. C. Pitt, J. C.
Beasley, and G. W. Witmer (Boca Raton, FL: CRC Press, 2017), 415.

43 four days after deployment: R. A. Garcia, et al., "Adaptation of an Artificial Bait
to an Automated Aerial Delivery System for Landscape-Scale Brown Treesnake
Suppression," *Biological Invasions* 23 (May 2021), https://doi.org/10.1007
/s10530-021-02567-8.

44 indistinguishable from Spam: Mary Roach, *Fuzz: When Nature Breaks the Law*
(New York: W. W. Norton & Company, 2021).

44 6 percent of the bait: S. R. Siers, et al., "In Situ Evaluation of an Automated
Aerial Bait Delivery System for Landscape-Scale Control of Invasive Brown
Treesnakes on Guam," in *Island Invasives: Scaling Up to Meet the Challenge*,
ed. C. R. Veitch, et al. (Gland, Switzerland: IUCN, 2019), 348–55.

45 four grams of acetaminophen: S. R. Siers, et al., "Evaluating Lethal Toxicant
Doses for the Largest Individuals of an Invasive Vertebrate Predator with
Indeterminate Growth," *Management of Biological Invasions* 12, no. 2 (June
2021), https://doi.org/10.3391/mbi.2021.12.2.17.

48 month in salt water: K. M. Hart, P. J. Schofield, and D. R. Gregoire,
"Experimentally Derived Salinity Tolerance of Hatchling Burmese Pythons
(*Python molurus bivittatus*) from the Everglades, Florida (USA)," *Journal of
Experimental Marine Biology and Ecology* 413 (February 2012), https://doi
.org/10.1016/j.jembe.2011.11.021.

49 Indian pythons: M. E. Hunter, et al., "Cytonuclear Discordance in the
Florida Everglades Invasive Burmese Python (*Python bivittatus*) Population
Reveals Possible Hybridization with the Indian Python (*P. molurus*)," *Ecology
and Evolution* 8 (September 2018), https://doi.org/10.1002/ece3.4423.

49 the "Judas" technique: B. J. Smith, et al., "Betrayal: Radio-Tagged Burmese
Pythons Reveal Locations of Conspecifics in Everglades National Park,"
Biological Invasions 18 (July 2016), https://doi.org/10.1007/s10530-016-
1211-5.

51 the Cobra Effect: J. Maheshwari, "Cobra Effect: The Law of Unintended
Consequences (Part 1)," Medium, February 13, 2019, https://medium
.com/@jayna.1989/cobra-effect-the-law-of-unintended-consequences-part-1
-d3e674f68400.

CHAPTER 3: A NEST OF MICE

60 full of gazelles: T. Dayan and D. Simberloff, "Natufian Gazelles: Proto-
Domestication Reconsidered," *Journal of Archaeological Science* 22 (September
1995), https://doi.org/10.1016/S0305-4403(95)80152-9.

62 house mouse was born: B. Brookshire, "How the House Mouse Tamed Itself," *Science News*, April 19, 2017, https://www.sciencenews.org/blog/scicurious/how -house-mouse-tamed-itself.

63 Delicious: Gilbert Smith, et al., "Human Follicular Mites: Ectoparasites Becoming Symbionts," *Molecular Biology and Evolution* 39, no. 6 (June 2022), https://doi.org/10.1093/molbev/msac125.

64 new, scary environments: J. A. Bravo, et al., "Ingestion of Lactobacillus Strain Regulates Emotional Behavior and Central Gaba Receptor Expression in a Mouse via the Vagus Nerve," *Proceedings of the Natioanl Academy of Sciences* 108 (August 2011), https://doi.org/10.1073/pnas.1102999108.

64 produce the same effect: J. R. Kelly, et al., "Lost in Translation? The Potential Psychobiotic Lactobacillus Rhamnosus (Jb-1) Fails to Modulate Stress or Cognitive Performance in Healthy Male Subjects," *Brain, Behavior, and Immunity* 61 (March 2017): 50–59, https://doi.org/10.1016/j.bbi.2016.11.018.

64 pharmacy in your gut: A. Alberdi, et al., "Do Vertebrate Gut Metagenomes Confer Rapid Ecological Adaptation?," *Trends in Ecology & Evolution* 31, no. 9 (September 2016): 689–99, https://doi.org/10.1016/j.tree.2016.06.008.

65 that likes darkness: K. M. Neufeld, et al., "Reduced Anxiety-like Behavior and Central Neurochemical Change in Germ-Free Mice," *Neurogastroenterology & Motility* 23, no. 3 (March 2011): 255–64, https://doi.org/10.1111/j.1365 -2982.2010.01620.x.

65 no response at all: K-A. M. Neufeld, et al., "Mouse Strain Affects Behavioral and Neuroendocrine Stress Responses Following Administration of Probiotic *Lactobacillus rhamnosus* JB-1 or Traditional Antidepressant Fluoxetine," *Frontiers in Neuroscience* 12 (May 2018): 294, https://doi.org/10.3389 /fnins.2018.00294.

67 solving the tasks: L. Vrbanec, et al., "Enhanced Problem-Solving Ability as an Adaptation to Urban Environments in House Mice," *Proceedings of the Royal Society of Sciences—Biological Sciences* 288, no. 1945, February 2021, http:// doi.org/10.1098/rspb.2020.2504.

67 poor country mice: V. Mazza and A. Guenther, "City Mice and Country Mice: Innovative Problem Solving in Rural and Urban Noncommensal Rodents," *Animal Behaviour* 172 (February 2021): 197–210, https://doi.org/10.1016/j .anbehav.2020.12.007.

67 In urban environments: Pizza Ka Yee Chow, Nicola S. Clayton, and Michael A. Steele, "Cognitive Performance of Wild Eastern Gray Squirrels (*Sciurus carolinensis*) in Rural and Urban, Native, and Non-native Environments," *Frontiers in Ecology and Evolution* 9 (February 2021), https:// doi.og10.3389/fevo.2021.615899.

68 the behavior spread: L. Lefebvre, "The Opening of Milk Bottles by Birds: Evidence for Accelerating Learning Rates, but Against the Wave-of-Advance Model of Cultural Transmission," *Behavioural Processes* 34, no. 1 (May 1995): 43–53, https://doi.org/10.1016/0376-6357(94)00051-H.

68 raccoon-proof bins: L. Cecco, "Raccoons v Toronto: How 'Trash Pandas' Conquered the City," *Guardian*, October 5, 2018, https://www.theguardian .com/world/2018/oct/05/canada-toronto-raccoons.

72 forty-five hundred years ago: E. Cortesi, et al., "Cultural Relationships Beyond the Iranian Plateau: The Helmand Civilization, Baluchistan and the Indus Valley in the 3rd Millennium BCE," *Paléorient* 34, no. 2 (January 2008): 5–35, http://doi.org/10.2307/41496521.

73 invention of William Hooker: W. C. Hooker, "Animal Trap," Patent No. 528,671, November 6, 1894.

73 the author of books: D. Drummond, *Nineteenth Century Mouse Traps Patented in the U.S.A.: An Illustrated Guide* (Galloway, OH: North American Trap Collectors Association, Inc., 2004).

74 *Queanbeyan Age* in 1871: S. M. Herald, "A Mouse Plague," *Queanbeyan Age*, June 15, 1871.

75 thirty-two million mice: "Mice Plague in Australia," *Nature* 129, no. 755 (May 1932), https://doi.org/10.1038/129755b0.

75 their hospital beds: N. Zhou, "Three Hospital Patients Bitten as Mouse Plague Sweeps Western NSW," *Guardian*, March 18, 2021, https://www.theguardian .com/australia-news/2021/mar/18/three-hospital-patients-bitten-by-mice-as -absolute-plague-sweeps-western-nsw.

75 with zinc phosphide: "'Follow the Instructions,' Customers Warned after Several Hospitalized by Rodent Bait," *Mudgee Guardian* (February 9, 2021), https://www.mudgeeguardian.com.au/story/7118595/residents-urged-to-be -cautious-with-mouse-bait-misuse-resulting-in-poisoning/.

75 as well as it does rats: V. Olmos and C. Magdalena López, "Brodifacoum Poisoning with Toxicokinetic Data," *Clinical Toxicology* 45, no. 5 (October 2008): 487–89, doi: 10.1080/15563650701354093.

78 as one unit: Reka K. Kelemen, Marwan Elkrewi, Anna K. Lindholm, and Beatriz Vicoso, "Novel Patterns of Expression and Recruitment of New Genes on the T-Haplotype, a Mouse Selfish Chromosome," *Proceedings of the Royal Society B* 289, no. 1968 (February, 2022): 20211985, https://doi.org/10.1098/rspb.2021.1985.

78 95 percent of the time: A. Manser, B. König, and A. K. Lindholm, "Polyandry Blocks Gene Drive in a Wild House Mouse Population," *Nature Communications* 11, no. 5590 (November 2020), https://doi.org/10.1038 /s41467-020-18967-8.

79 problem is over: J. Godwin, et al., "Rodent Gene Drives for Conservation: Opportunities and Data Needs," *Proceedings of the Royal Society—Biological Sciences* 286 (November 2019), http://doi.org/10.1098/rspb.2019.1606.

79 mice in a natural environment: J. N. Runge, "Selfish Migrants: How a Meiotic Driver Is Selected to Increase Dispersal," *Journal of Evolutionary Biology* 35, no 4 (April 2022): 621–32, https://doi.org/10.1111/jeb.13989.

80 won't even mate: A. Manser, et al., "Female House Mice Avoid Fertilization by T Haplotype Incompatible Males in a Mate Choice Experiment," *Journal of*

Evolutionary Biology 28, no. 3 (January 2015): 54–64, https://doi.org/10.1111 /jeb.12525.

80 it wasn't that simple: A. Manser, et al., "Controlling Invasive Rodents via Synthetic Gene Drive and the Role of Polyandry," *Proceedings of the Royal Society—Biological Sciences* 286 (August 2019), https://doi.org/10.1098 /rspb.2019.0852.

82 50 percent to 72 percent: H. A. Grunwald, et al., "Super-Mendelian Inheritance Mediated by CRISPR–Cas9 in the Female Mouse Germline," *Nature* 566 (January 2019): 105–9, https://doi.org/10.1038/s41586-019-0875-2/.

84 dogs and sheep: "History of Blood Transfusion," American Red Cross, https://www .redcrossblood.org/donate-blood/blood-donation-process/what-happens-to-donated -blood/blood-transfusions/history-blood-transfusion.html, accessed May 18, 2022.

84 "Do you like mice?": C. C. Little, "A New Deal for Mice," *Scientific American*, January 1, 1935.

84 to form the Jackson Laboratory: Leila McNeill, "The History of Breeding Mice for Science Begins with a Woman in a Barn," *Smithsonian Magazine*, March 20, 2018.

84 "nature and cure of cancer": Little, "A New Deal for Mice."

85 says Karen Rader: K. Rader, *Making Mice: Standardizing Animals for American Biomedical Research, 1900–1955* (Princeton, NJ: Princeton University Press, 2004).

85 twelve thousand strains of mice: "Fast Facts," Jackson Laboratory, https://www .jax.org/about-us/fast-facts, accessed May 18, 2022.

85 three million mice: "The World's Favourite Lab Animal Has Been Found Wanting, but There Are New Twists in the Mouse's Tale," *The Economist*, December 24, 2016, https://www.economist.com/christmas-specials/2016/12/24 /the-worlds-favourite-lab-animal-has-been-found-wanting-but-there-are-new -twists-in-the-mouses-tale.

CHAPTER 4: A DROPPING OF PIGEONS

89 found modern journalism: "The Long History of Speed at Reuters," Reuters, October 21, 2020, https://www.reuters.com/article/rpb-historyofspeed -idUSKBN2761XC.

89 and South Asia: Michael D. Shapiro and Eric T. Domyan, "Domestic Pigeons," *Current Biology* 23, no. 8 (April 2013): PR302-R303, https://doi.org/10.1016 /j.cub.2013.01.063.

90 earliest domesticated birds: D. S. Farner, et al., ed., *Avian Biology VI* (New York: Academic Press, 1982).

90 five thousand years ago: C. A. Driscoll, et al., "From Wild Animals to Domestic Pets, an Evolutionary View of Domestication," *Proceedings of the National Academy of Sciences*, 106 (June 2009): 9971–78, https://doi.org/10.1073 /pnas.0901586106.

90   is the messenger: Andrew D. Blechman, *Pigeons: The Fascinating Saga of the World's Most Revered and Reviled Bird* (New York: Grove Press, 2006), 13.

90   the pigeon fancy: C. Darwin, *On the Origin of Species: A Facsimilie of the First Edition* (Cambridge: Harvard University Press, 1964).

91   feral pigeon populations do: J. Jokimäki and J. Suhonen, "Distribution and Habitat Selection of Wintering Birds in Urban Environments," *Landscape and Urban Planning* 39, no. 4 (January 1998): 253–63, https://doi.org/10.1016 /S0169-2046(97)00089-3.

91   June 22, 1966: "Hoving Calls a Meeting to Plan for Restoration of Bryant Park; Cleanup Is Urged for Bryant Park," *New York Times*, June 22, 1966, https://www .nytimes.com/1966/06/22/archives/hoving-calls-a-meeting-to-plan-for -restoration-of-bryant-park.html.

91   1851 to 2006: C. Jerolmack, "How Pigeons Became Rats: The Cultural-Spatial Logic of Problem Animals," *Social Problems* 55, no. 1 (February 2008): 72–94, https://doi.org/10.1525/sp.2008.55.1.72.

91   birds of spreading disease: Jerolmack, "How Pigeons Became Rats."

91   dovecotes of the rich: C. Humphries, *Superdove: How the Pigeon Took Manhattan . . . and the World* (Washington, DC: Smithsonian, 2008).

92   in ancient Persia: Humphries, *Superdove*.

92   two whole hours: "Reuters: A Brief History," *Guardian*, May 4, 2007, https:// www.theguardian.com/media/2007/may/04/reuters.pressandpublishing.

93   an urban environment: D. Ducatez, et al., "Ecological Generalism and Behavioural Innovation in Birds: Technical Intelligence or the Simple Incorporation of New Foods?," *Journal of Animal Ecology* 84 (June 2014): 79–89, https://doi.org/10.1111/1365-2656.12255.

94   urban birds are better: J. Audet, "The Town Bird and the Country Bird: Problem Solving and Immunocompetence Vary with Urbanization," *Behavioral Ecology* 27, no. 2 (March–April 2016): 637–44, https://doi.org/10.1093/beheco/arv201.

95   it was thirteen rabbits: "How European Rabbits Took Over Australia," *National Geographic*, https://www.nationalgeographic.org/article/how-european-rabbits -took-over-australia/, accessed May 18, 2022.

95   rabbit-proof fences: "State Barrier Fence Overview," Department of Primary Industries and Regional Development, Government of Western Australia, updated May 4, 2022, https://www.agric.wa.gov.au/invasive-species/state-barrier -fence-overview.

98   infected with "ornithosis": Jerolmack, "How Pigeons Became Rats."

98   also called psittacosis: Jerolmack, "How Pigeons Became Rats."

98   a health menace: Jerolmack, "How Pigeons Became Rats."

98   DO NOT FEED PIGEONS: "Health Board Bids City Abolish Pigeon-Feeding Areas in Parks," *New York Times*, October 23, 1963, https://www.nytimes.com/1963 /10/23/archives/health-board-bids-city-abolish-pigeonfeeding-areas-in-parks .html.

99   around the globe: A. P. Litvintseva, et al., "Evidence That the Human Pathogenic Fungus Cryptococcus Neoformans Var. Grubii May Have Evolved

in Africa," *PLOS ONE* 11, no. 6 (May 2011), doi: 10.1371/journal
.pone.0019688.

100 70 percent of Australia: T. Strive and T. E. Cox, "Lethal Biological Control of
Rabbits—The Most Powerful Tools for Landscape-Scale Mitigation of Rabbit
Impacts in Australia," *Australian Zoologist* 40, no. 1 (2019): 118–29, https://doi
.org/10.7882/AZ.2019.016.

100 into the bush in the late 1880s: D. Peacock and I. Abbott, "The Mongoose in
Australia: Failed Introduction of a Biological Control Agent," *Australian Journal
of Zoology* 58, no. 4 (November 2010): 205–27, http://doi.org/ 10.1071
/ZO10043.

101 method to kill bunnies: "Extermination of Rabbits," *Sydney Morning Herald*,
September 7, 1867, https://trove.nla.gov.au/newspaper/article/13649462.

101 with chicken cholera: J. F. Prescott, review of *Pasteur's Gambit: Louis Pasteur,
the Australasian Rabbit Plague and a Ten Million Dollar Prize*, *Veterinary
Microbiology* 149 (May 2011): 3N4, http://doi.org/ 10.1016/j.vetmic
.2010.12.019.

101 *Myxoma* took off: T. Strive, "Lethal Biological Control of Rabbits," *Australian
Zoologist* 40, no. 1 (2019): 118–28.

101 PR and humane disaster: P. J. Kerr, R. N. Hall, and T. Strive, "Viruses for
Landscape-Scale Therapy: Biological Control of Rabbits in Australia," in
Viruses as Therapeutics: Methods and Protocols, ed. A. R. Lucas (New York,
Humana Press, 2021).

101 imported from Germany: Kerr, Hall, and Strive, "Viruses for Landscape-Scale
Therapy."

102 months after death: Kerr, Hall, and Strive, "Viruses for Landscape-Scale
Therapy."

102 around the outbreak: Kerr, Hall, and Strive, "Viruses for Landscape-Scale
Therapy."

102 pest control in 1996: Kerr, Hall, and Strive, "Viruses for Landscape-Scale
Therapy."

102 another, deadlier strain: D. S. L. Ramsey, et al., "Emerging RHDV2 Suppresses
the Impact of Endemic and Novel Strains of RHDV on Wild Rabbit
Populations," *Journal of Applied Ecology* 57, no. 3 (March 2020): 630–41,
https://doi.org/10.1111/1365-2664.13548.

103 pounds of waste per year: D. H. R. Spennemann and M. J. Watson,
"Experimental Studies on the Impact of Bird Excreta on Architectural Metals,"
APT Bulletin: Journal of Preservation Technology 49, no. 1 (2018): 19–28, https://
www.jstor.org/stable/26452201.

103 "aesthetically unpleasing": D. H. R. Spennemann, M. Pike, and M. J. Watson,
"Effects of Acid Pigeon Excreta on Building Conservation," *International
Journal of Building Pathology* 35, no. 1 (April 2017), https://doi.org/10.1108
/IJBPA-09-2016-0023.

104 Sacré-Coeur and Trafalgar Square: "How Does Acid Precipitation Affect
Marble and Limestone Buildings?," United States Geological Survey, https://

www.usgs.gov/faqs/how-does-acid-precipitation-affect-marble-and-limestone
-buildings?qt-news_science_products=0#, accessed May 18, 2022.

104 pH of 7.4: Spennemann, Pike, and Watson, "Effects of Acid Pigeon Excreta."

104 copper or bronze: E. Bernardi, et al., "The Effect of Uric Acid on Outdoor
Copper and Bronze," *Science of the Total Environment* 407, no. 7 (March
2009): 2383–89, https://doi.org/10.1016/j.scitotenv.2008.12.014.

104 care much about pigeons: D. H. R. Spennemann, M. Pike, and M. J. Watson,
"Bird Impacts on Heritage Buildings: Australian Practitioners' Perspectives
and Experiences," *Journal of Cultural Heritage Management and Sustainable
Development* 8, no. 1 (January 2018): 62–75, https://doi.org/10.1108/JCHMSD
-07-2016-0042.

105 can land anyway: M. Conover, *Resolving Human-Wildlife Conflicts: The
Science of Wildlife Damage Management* (Boca Raton, FL: CRC Press, 2001).

105 pigeon to land: Conover, *Resolving Human-Wildlife Conflicts.*

105 pack of fifteen: "Optical Bird Gel Repellant," BirdBusters, https://www
.birdbusters.com/shop/Optical-Bird-Gel-Repellent.html, accessed May 18, 2022.

106 Mikhail tracked down: A. Mikhail, *The Animal in Ottoman Egypt* (Oxford:
Oxford University Press, 2016).

107 dog's saliva is impure: "Fatwa No: 335128: Impurity of Dogs—Command to
Kill Black Dogs," Fatwa, October 22, 2016, https://www.islamweb.net/en
/fatwa/335128/impurity-of-dogs-command-to-kill-black-dogs.

107 at their ease: "Fatwa No: 335128."

107 done to a person: S. A. Rahman, "Religion and Animal Welfare—An Islamic
Perspective," *Animals (Basel)* 7, no. 2 (February 2017), http://doi.org/ 10.3390
/ani7020011.

107 they ate garbage: Mikhail, *The Animal in Ottoman Egypt.*

108 littered the streets: Mikhail, *The Animal in Ottoman Egypt.*

110 scare the pigeons away: L. Hornack, "London's Pigeon Problem Has a Simple
Solution: A Hawk," *The World*, March 15, 2017, https://theworld.org
/stories/2017-03-15/londons-pigeon-problem-has-simple-solution-hawk.

110 birdseed to tourists: C. Jerolmack, *The Global Pigeon* (Chicago: University of
Chicago Press, 2013).

111 writes in his book: Jerolmack, *The Global Pigeon.*

111 "put in a home": Jerolmack, *The Global Pigeon.*

CHAPTER 5: A MEMORY OF ELEPHANTS

118 adoption fee: "Adopt an Orphan," Sheldrick Wildlife Trust, https://www
.sheldrickwildlifetrust.org/orphans, accessed May 19, 2022.

119 terrifying game animals: Stephanie Hanes, *White Man's Game: Saving Animals,
Rebuilding Eden, and Other Myths of Conservation in Africa* (New York:
Metropolitan Books, 2017), 101.

119 by the millions: J. R. Poulsen, , C. Rosin, A. Meier, E. Mills, C. L. Nuñez, S.
E. Koerner, E. Blanchard, J. Callejas, S. Moore, and M. Sowers, "Ecological

Consequences of Forest Elephant Declines for Afrotropical Forests," *Conservation Biology* 32, no. 3 (June 2018): 559–67, https://doi.org/10.1111/cobi.13035.

121 2,398 people died in India: PTI, "Elephants Killed over 2,300 People in Last Five Years: Elephant Ministry," *The Hindu*, June 28, 2019, https://www.thehindu.com/sci-tech/energy-and-environment/elephants-killed-over-2300-people-in-last-five-years-environment-ministry/article28208456.ece.

121 450 elephants died, too: PTI, "Elephants Killed over 2,300 People."

121 in the United States, it's news: Normvance, "Bear Put Down after Trapping Family in Home for 45 Minutes," *Pagosa Springs Journal*, September 16, 2021, https://pagosasprings.com/bear-put-down-after-trapping-family-in-home-for-45-minutes/.

121 In a 2021 map: E. Di Minin, et al., "A Pan-African Spatial Assessment of Human Conflicts with Lions and Elephants," *Nature Communications* 12, no. 2978 (May 2021), https://doi.org/10.1038/s41467-021-23283-w.

122 parks and reserves in East Africa: Di Minin, et al., "A Pan-African Spatial Assessment."

122 and employs about one million people: Reuters Staff, "COVID Slashes Kenyan Tourism Revenues by $1 Billion," Reuters, December 2, 2020, https://www.reuters.com/article/health-coronavirus-kenya-tourism-idUSL8N2II4DE.

123 at least a generation: Jon T. Coleman, *Vicious: Wolves and Men in America* (New Haven: Yale University Press, Lamar Series in Western History, 2006).

123 horrible, hairy criminals: Coleman, *Vicious: Wolves and Men in America*.

123 the wolfish sinner: Coleman, *Vicious: Wolves and Men in America*.

123 "punished for living": Coleman, *Vicious: Wolves and Men in America*.

124 "an entire species": Coleman, *Vicious: Wolves and Men in America*.

124 lives per year: G. Bombieri, J. Naves, V. Penteriani, et al. "Brown Bear Attacks on Humans: a Worldwide Perspective," *Scientific Reports* 9, no. 8573 (June 2019), https://doi.org/10.1038/s41598-019-44341-w.

124 Chignik in Alaska: S. Woodham, "Wolf May Have Killed Teacher Near Chignik Lake," *Anchorage Daily News*, March 20, 2010, https://www.adn.com/alaska-beat/article/wolf-may-have-killed-teacher-near-chignik-lake/2010/03/10/.

124 2,040 cattle: USDA, "Cattle and Calves Death Loss in the United States Due to Predator and Nonpredator Causes, 2015," APHIS, VS, NAHMS, December 20, 2017.

124 mostly to disease: USDA, "Cattle and Calves."

125 *The Predator Paradox*: J. Shivik, *The Predator Paradox: Ending the War with Wolves, Bears, Cougars, and Coyotes* (Boston: Beacon Press, 2014).

126 notes in his book: Joseph Marshall III, *On Behalf of the Wolf and the First Peoples* (Santa Fe: Museum of New Mexico Press, 1995).

127 When a male elephant: Anthony J. Hall-Martin and L. A. van der Walt, "Plasma Testosterone Levels in Relation to Musth in the Male African Elephant," *Koedoe* 27 (December 1984): 147–49, https://doi.org/10.4102/koedoe.v27i1.561.

129 function of the ecosystem: Michiel P. Veldhuis, Mark E. Ritchie, Joseph O. Ogutu, et al., "Cross-Boundary Human Impacts Compromise the Serengeti-Mara Ecosystem," *Science* 363, no. 64636 (March 2019): 1424–28, https://doi .org/10.1126/science.aav0564.

129 leave the elephants in crisis: Connor J. Cavanagh, Teklehaymanot Welde-michel, and Tor A. Benjaminsen, "Gentrifying the African Landscape: The Performance and Powers of for-Profit Conservation on Southern Kenya's Con-servancy Frontier," *Annals of the American Association of Geographers* 110, no. 5 (March 2020): 1594–612, https://doi.org/10.1080/24694452.2020.1723398.

130 Electric fences are going up: J. Shaffer, "Human-Elephant Conflict: A Review of Current Management Strategies and Future Directions," *Frontiers in Ecology and Evolution* 6 (January 2019), https://doi.org/10.3389/fevo.2018.00235.

130 short-circuit the wires: Shaffer, "Human-Elephant Conflict."

130 fences to kill: T. Kalam, et al., "Lethal Fence Electrocution: A Major Threat to Asian Elephants in Assam, India," *Tropical Conservation Science* 11 (December 2018), https://doi.org/10.1177/1940082918817283.

130 Pliny the Elder: E. J. Christie, "The Idea of an Elephant: Ælfric of Eynsham, Epistemology, and the Absent Animals of Anglo-Saxon England," *Neophilologus* 98 (October 2013): 465–79, https://doi.org/10.1007/s11061-013 -9374-0.

130 had bees in them: L. E. King, "African Elephants Run from the Sound of Disturbed Bees," *Current Biology* 17, no. 19 (2007): R832–33, https://doi .org/10.1016/j.cub.2007.07.038.

130 they ran: King, "African Elephants Run."

131 86 percent: L. E. King, et al., "Beehive Fence Deters Crop-Raiding Elephants," *African Journal of Ecology* 47, no. 2 (June 2009): 131–37, https://doi .org/10.1111/j.1365-2028.2009.01114.x.

131 twenty countries and counting: "Our Beehive Fence Design," Elephants and Bees Project, Save the Elephants, https://elephantsandbees.com/beehive-fence/, accessed May 18, 2022.

131 full of bees: King, et al., "Beehive Fence Deters Crop-Raiding Elephants."

131 $2,500 per year: G. Rapsomanikis, "The Economic Lives of Smallholder Farmers," Food and Agriculture Organization of the United Nations, 2015.

132 an elephant repellant: T. Chapman, "Brewing Smelly Elephant Repellant," Elephants and Bees Project, Save the Elephants, December 17, 2019, https:// elephantsandbees.com/brewing-smelly-elephant-repellent/.

134 often afraid of bees: Renaud Hecklé, Pete Smith, Jennie I. Macdiarmid, Ewan Campbell, Pamela Abbott, "Beekeeping Adoption: a Case Study of Three Small-holder Farming Communities in Baringo County, Kenya," *Journal of Agriculture and Rural Development in the Tropics and Subtropics* 119, no. 1 (2018).

135 a geo-fence: J. Wall, et al., "Novel Opportunities for Wildlife Conservation and Research with Real-Time Monitoring," *Ecological Applications* 24, no. 4 (June 2014): 593–601, http://doi.org/10.1890/13-1971.1.

135 fifty "flagship" elephants: "Monitor," Mara Elephant Project, https://
     maraelephantproject.org/our-approach/monitor/, accessed May 18, 2022.

136 three quadcopter drones: N. Hahn, et al., "Unmanned Aerial Vehicles Miti-
     gate Human–Elephant Conflict on the Borders of Tanzanian Parks: A Case
     Study," *Oryx* 51, no. 3 (November 2016): 513–16, doi: 10.1017
     /S0030605316000946.

136 the propeller blades: "Protect," Mara Elephant Project, https://
     maraelephantproject.org/our-approach/protect/, accessed May 18, 2022.

137 and life in prison: E. Atienza, "Fact Check: Did Kenya Introduce the Death
     Penalty for Wildlife Poachers?," Checkyourfact.com, December 25, 2019,
     https://checkyourfact.com/2019/12/25/fact-check-kenya-introduce-death
     -penalty-law-poachers-elephants-rhinos/.

140 exceptionally good luck: "The Elephant's Placenta and the Lucky Brothers,"
     Lion Guardians, June 9, 2017, http://lionguardians.org/the-elephants-placenta
     -and-the-lucky-brothers/.

140 to receive blessings: J. Kioke, et al., "Maasai People and Elephants: Values and
     Perceptions," *Indian Journal of Traditional Knowledge* 14, no. 1 (January 2015):
     13–19.

CHAPTER 6: A NUISANCE OF CATS

145 one of her sisters: B. Brookshire, "I Spent 5 Months Trying to Coax a Cat from
     My Ceiling," *Atlantic*, August 19, 2021, https://www.theatlantic.com/science
     /archive/2021/08/ceiling-cat-meme-came-live-my-house/619832/.

146 one to four billion birds and six to twenty-two billion mammals: Loss, Will, and
     Marra, "The Impact of Free-Ranging Domestic Cats."

146 At least 63 species: F. Medina, et al., "A Global Review of the Impacts of
     Invasive Cats on Island Endangered Vertebrates," *Global Change Biology* 17
     (November 2011), https://doi.org/10.1111/j.1365-2486.2011.02464.x.

146 threaten 430 more: T. S. Doherty, et al., "Invasive Predators and Global
     Biodiversity Loss," *Proceedings of the National Academy of Sciences* 40 (October
     2015), https://doi.org/10.1073/pnas.1602480113.

146 86 percent of the species extinctions: D. R. Spatz, et al., "Globally Threatened
     Vertebrates on Islands with Invasive Species," *Science Advances* 3 (October
     2017), doi:10.1126/sciadv.1603080.

147 around ten thousand years ago: C. A. Driscoll, et al., "The Near Eastern Origin
     of Cat Domestication," *Science* 317 (July 2007), https://doi.org/10.1126
     /science.1139518.

147 basically called cats "meow": J. Bradshaw, *Cat Sense: How the New Feline Science
     Can Make You a Better Friend to Your Pet* (New York: Basic Books, 2013).

147 they were being depicted: J. A. Serpall, "Domestication and History of the Cat,"
     in *The Domestic Cat: The Biology of its Behaviour* (Cambridge: Cambridge
     University Press, 2013), 89.

147 were caring for kitty injuries: A. Haruda, et al., "The Earliest Domestic Cat on the Silk Road," *Scientific Reports* 10 (2020), https://doi.org/10.1038/s41598-020 -67798-6.

148 in 1998, their traps came up empty: E. Vázquez-Domínguez, G. Ceballos, and J. Cruzado, "Extirpation of an Insular Subspecies by a Single Introduced Cat: The Case of the Endemic Deer Mouse *Peromyscus guardia* on Estanque Island, Mexico," *Oryx* 38 (August 2004), doi: 10.1017 /S0030605304000602.

148 a cat, and one hundred of her bowel movements: E. Mellink, G. Ceballos, and J. Luévano, "Population Demise and Extinction Threat of the Angel De La Guarda Deer Mouse (Peromyscus Guardia)," *Biological Conservation* 108 (November 2002), https://doi.org/10.1016/S0006-3207(02)00095-2.

148 93 percent of the scats: Vázquez-Domínguez, Ceballos, and Cruzado, "Extirpation of an Insular Subspecies."

148 The last known example died: Vázquez-Domínguez, Ceballos, and Cruzado, "Extirpation of an Insular Subspecies."

149 In just a year, the bird was gone: R. Galbreath and D. Brown, "The Tale of the Lighthouse-Keeper's Cat: Discovery and Extinction of the Stephens Island Wren (*Traversia lyalli*)," *Notornis* 51 (2004).

149 cats disappeared from the island: Galbreath and Brown, "The Tale of the Lighthouse-Keeper's Cat."

150 The analog towers: "Hawaii TV Stations to Go Digital One Month before National DTVTransition," *Hawaii News Now*, October 15, 2008.

151 In a survey conducted: A. F. Raine, et al., "Managing the Effects of Introduced Predators on Hawaiian Endangered Seabirds," *Journal of Wildlife Management* 84 (April 2020), https://doi.org/10.1002/jwmg.21824.

151 Half of those were due to rats: Raine, "Managing the Effects of Introduced Predators."

151 headed for extinction within: Raine, "Managing the Effects of Introduced Predators."

153 2.8 million feral cats: House of Representatives Standing Committee on the Environment and Energy, "Tackling the Feral Cat Pandemic: A Plan to Save Australian Wildlife: Report of the Inquiry into the Problem of Feral and Domestic Cats in Australia," Commonwealth of Australia, 2020, https://www .aph.gov.au/Parliamentary_Business/Committees/House/Environment_and _Energy/Feralanddomesticcats/Report.

153 30 million homeless felines: A. N. Rowan, T. Kartal, and J. Hadidian, "Cat Demographics & Impact on Wildlife in the USA, the UK, Australia and New Zealand: Facts and Values," *Journal of Applied Animal Ethics Research* 2 (2019), https://doi.org/10.1163/25889567-12340013.

153 extinction of twenty-five Australian mammal species: House of Representatives Standing Committee on the Environment and Energy, "Tackling the Feral Cat Pandemic."

153 prevented the extinction of thirteen mammal taxa: S. Legge, et al., "Australia Must Control Its Killer Cat Problem. A Major New Report Explains How, but Doesn't Go Far Enough," The Conversation, February 9, 2021.

155 2005 paper looking at TNR programs: P. Foley, et al., "Analysis of the Impact of Trap-Neuter-Return Programs on Populations of Feral Cats," *Journal of the American Veterinary Medical Association* 227 (December 2005), https://doi .org/10.2460/javma.2005.227.1775.

155 if you could TNR: Foley, et al., "Analysis of the Impact of Trap-Neuter-Return Programs."

155 bred even more: Idit Gunther, et al., "Reduction of Free-Roaming Cat Population Requires High-Intensity Neutering in Spatial Contiguity to Mitigate Compensatory Effects," *Proceedings of the National Academy of Sciences* 119, no. 15 (April 2022): e2119000119, https://doi.org/10.1073/pnas.2119000119.

156 by Daniel Spehar and Peter Wolf: D. D. Spehar and P. J. Wolf, "Back to School: An Updated Evaluation of the Effectiveness of a Long-Term Trap-Neuter-Return Program on a University's Free-Roaming Cat Population," *Animals* 9 (2019), https://doi.org/10.3390/ani9100768.

156 there were only ten cats left: Spehar and Wolf, "Back to School."

156 In most successful studies: K. Tan, J. Rand, and J. Morton, "Trap-Neuter-Return Activities in Urban Stray Cat Colonies in Australia," *Animals* 7 (2017), https:// doi.org/10.3390/ani7060046. See also: Spehar and Wolf, "Back to School."

158 "climate refugees": M. C. Urban, "Climate-Tracking Species Are Not Invasive," *Nature Climate Change* 10 (May 2020): 382–84, https://doi.org/10.1038 /s41558-020-0770-8.

158 "Instead of a paradigm": Marris, *Wild Souls*, 173.

159 estimated in 2013 that cats kill: Loss, Will, and Marra, "The Impact of Free-Ranging Domestic Cats"; S. Loss, T. Will, and P. Marra, "Correction: Corrigendum: The Impact of Free-Ranging Domestic Cats on Wildlife of the United States," *Nature Communications* 4 (2013), https://doi.org/10.1038/ncomms3961.

159 925 GPS collars: R. Kays, et al., "The Small Home Ranges and Large Local Ecological Impacts of Pet Cats," *Animal Conservation* 23 (October 2020), https://doi.org/10.1111/acv.12563.

160 They're about forty-one calories each: F. B. Golley, "Energy Dynamics of a Food Chain of an Old-Field Community," *Ecological Monographs* 30 (April 1960), https://doi.org/10.2307/1948551.

160 delicious, delicious meat: Martina Cecchetti, et al., "Drivers and Facilitators of Hunting Behaviour in Domestic Cats and Options for Management," *Mammal Review* 51, no. 3 (July 2021): 307–22, https://doi.org/10.1111 /mam.12230.

160 Fancy Feast throughout history: Cecchetti et al., "Drivers and Facilitators of Hunting Behaviour."

160 keep the hunting hustle going: Cecchetti et al., "Drivers and Facilitators of Hunting Behaviour."

161 rats and cats just avoided each other: M. H. Parsons, et al., "Temporal and Space-Use Changes by Rats in Response to Predation by Feral Cats in an Urban Ecosystem," *Frontiers in Ecology and Evolution* (September 2018), https://doi .org/10.3389/fevo.2018.00146.

161 surveyed 219 feline households: S. L. Crowley, M. Cecchetti, and R. A. McDonald, "Diverse Perspectives of Cat Owners Indicate Barriers to and Opportunities for Managing Cat Predation of Wildlife," *Frontiers in Ecology and the Environment* 18 (December 2020), https://doi.org/10.1002 /fee.2254.

162 41 percent of the population: W. L. Linklater, et al., "Prioritizing Cat-Owner Behaviors for a Campaign to Reduce Wildlife Depredation," *Conservation Science and Practice* 5 (May 2019), https://doi.org/10.1111/csp2.29.

163 apparently without any hint of irony: G. Williams, "Caught Cats Put to Good Use Catching Rabbits in Queenstown," *Otago Daily Times*, July 18, 2021.

163 study that Linklater and his colleagues: Linklater, et al., "Prioritizing Cat-Owner Behaviors."

165 cat-owning beliefs into five groups: Crowley, Cecchetti, and McDonald, "Diverse Perspectives of Cat Owners."

166 what they might be willing to do: W. L. Linklater, et al., "Prioritizing Cat-Owner Behaviors."

167 decreased the prey brought home: M. Cecchetti, et al., "Provision of High Meat Content Food and Object Play Reduce Predation of Wild Animals by Domestic Cats *Felis catus*," *Current Biology* 31 (2021), https://doi.org/10.1016 /j.cub.2020.12.044.

167 cultural models that different groups use: K. M. Leong, A. R. Gramza, and C. A. Lepczyk, "Understanding Conflicting Cultural Models of Outdoor Cats to Overcome Conservation Impasse," *Conservation Biology* 34 (October 2020), https://doi.org/10.1111/cobi.13530.

CHAPTER 7: A BAND OF COYOTES

177 attack our kids: Jesse O'Neill, "Coyote Attacks Toddler on California Beach," *New York Post*, April 30, 2022, https://nypost.com/2022/04/30/huntington -beach-coyote-attack-injures-toddler-in-california/.

177 Manhattan's Central Park: J. Salo, "Coyote Spotted in Central Park," *New York Post*, February 8, 2021, https://nypost.com/2021/02/08/coyote-spotted-in-central -park/.

178 Rancho La Brea in Los Angeles: "Mammal Collections," La Brea Tar Pits & Museum, https://tarpits.org/research-collections/tar-pits-collections/mammal -collections, accessed May 28, 2022.

178 *Coyote America*: Dan Flores, *Coyote America: A Natural and Supernatural History* (New York: Basic Books, 2016).

178 Steve Pavlik: Steve Pavlik, *Navajo and the Animal People: Native American Traditional Ecological Knowledge and Ethnozoology* (Golden, CO: Fulcrum Publishing, 2014).

178 First Man and First Woman: Pavlik, *Navajo and the Animal People.*

178 across the sky: Pavlik, *Navajo and the Animal People.*

178 death as well as birth: Pavlik, *Navajo and the Animal People.*

179 burned-up wood for his trouble: Pavlik, *Navajo and the Animal People.*

179 without much trouble: K. M. Berger, et al., "Indirect Effects and Traditional Trophic Cascades: A Test Involving Wolves, Coyotes and Pronghorn," *Ecology* 89, no. 3 (March 2008): 818–28, https://doi .org/10.1890/07-0193.1.

180 pack for protection: Berger, et al., "Indirect Effects and Traditional Trophic Cascades."

180 They're bigger: J. G. Way, "A Comparison of Body Mass of *Canis latrans* (Coyotes) between Eastern and Western North America," *Northeastern Naturalist* 14, no. 1 (March 2007): 111–24, https://doi.org/10.1656/1092 -6194(2007)14[111:ACOBMO]2.0.CO;2.

181 it's hard to know: Flores, *Coyote America.*

181 start breeding slightly earlier: J. C. Kligo, et al., "Reproductive Characteristics of a Coyote Population Before and During Exploitation," *Journal of Wildlife Management* 81, no. 6 (November 2017): 1386–93, https://doi.org/10.1002 /jwmg.21329.

181 for the rest: Kligo, et al., "Reproductive Characteristics of a Coyote Population."

181 were moving in: Kligo, et al., "Reproductive Characteristics of a Coyote Population."

182 it was the yard: S. M. Alexander and D. L. Draper, "The Rules We Make That Coyotes Break," *Contemporary Social Science* 16, no. 1 (May 2019): 127–39, http://doi.org/10.1080/21582041.2019.1616108.

182 always be trespassing: Alexander and Draper, "The Rules We Make That Coyotes Break."

183 more natural landscape: S. M. Alexander and D. L. Draper, "Worldviews and Coexistence with Coyotes," in *Human Wildlife Interactions: Turning Conflict into Coexistence*, ed. Beatrice Frank, Jenny A. Glikman, and Silvio Marchini (Cambridge: Cambridge University Press, 2019).

184 compared to the outskirts: M. Fidino, et al., "Landscape-Scale Differences among Cities Alter Common Species' Responses to Urbanization," *Ecological Applications* 31, no. 2 (March 2021): e02253, https://doi.org /10.1002/eap.2253.

185 don't want to: E. H. Ellington and S. D. Gehrt,"Behavioral Responses by an Apex Predator to Urbanization," *Behavioral Ecology* 30, no. 3 (May/June 2019): 821–29, https://doi.org/10.1093/beheco/arz019.

185 middle of downtown: Ellington and Gehrt, "Behavioral Responses by an Apex Predator."

185 posts and any comment threads: S. Altrudi and C. Kelty, "Animals, Angelenos and the Arbitrary: Analyzing Human-Wildlife Entanglement in Los Angeles," personal communication, 2021.

185 "keep your pets inside!": Altrudi and Kelty, "Animals, Angelenos and the Arbitrary."

185 make people afraid: Altrudi and Kelty, "Animals, Angelenos and the Arbitrary."

186 make the news: B. Erickson, "Lake Highlands Residents, Authorities Tell Different Stories About Coyote That Attacked Toddler," *Dallas* magazine, May 4, 2022, https://www.dmagazine.com/frontburner/2022/05/coyote-attack-dallas -lake-highlands/.

187 can startle them: N. J. Lance, et al., "Biological, Technical, and Social Aspects of Applying Electrified Fladry for Livestock Protection from Wolves (*Canis lupus*)," *Wildlife Research* 37, no. 8 (January 2011), http://doi.org/ 10.1071 /WR10022.

187 Starting in 2000, scientists: M. Musiani, et al., "Wolf Depredation Trends and the Use of Fladry Barriers to Protect Livestock in Western North America," *Conservation Biology* 17, no. 6 (December 2003): 1538–47, https://doi .org/10.1111/j.1523-1739.2003.00063.x.

188 it looked too small: J. Young, et al., "Mind the Gap: Experimental Tests to Improve Efficacy of Fladry for Nonlethal Management of Coyotes," *Wildlife Society Bulletin* 43, no. 9 (June 2019), http://doi.org/10.1002/wsb.970.

188 experimented with turbo-fladry: Lance, et al., "Biological, Technical, and Social Aspects of Applying Electrified Fladry."

190 grant applications in 2017: "Agriculture & Wildlife Protection Program," Benton County, OR., https://bentonawpp.wordpress.com/home/, accessed May 19, 2022.

191 little penguins of Middle Island: E. Nobel, "Maremma Sheepdog and Little Penguin Protector Retires after Nine Years on Middle Island," ABC News, October 16, 2019, https://www.abc.net.au/news/2019-10-17/middle-island -penguin-protector-oddball-maremma-retires/11607662.

194 as seen on *Shark Tank*!: CoyoteVest, https://www.coyotevest.com/products /coyotevest, accessed May 19, 2022.

195 black-and-white prey did: Caitlin Fay, "Aposematic Variation and the Evolution of Warning Coloration in Mammals" (master's thesis, California State University, Long Beach, 2016).

195 a warning signal: Fay, "Aposematic Variation," 27.

196 outdoors every day: Kathy Vo, speech at the Society of Integrative and Comparative Biology, 2021.

196 run-ins with coyotes: Vo, speech at the Society of Integrative and Comparative Biology.

196 encounter a coyote: "Take Action: Coexisting with Coyotes," Santa Monica Mountains National Recreation Area, National Park Service, https://www.nps .gov/samo/learn/management/support-coyotes.htm, accessed May 19, 2022.

196 In a small study: J. K. Young, E. Hammill, and S. W. Breck, "Interactions with Humans Shape Coyote Responses to Hazing," *Scientific Reports* 9 (December 2019), https://doi.org/10.1038/s41598-019-56524-6.
196 more hazing effort: Young, Hammill, and Breck, "Interactions with Humans."
197 while she was on tour: "Coyotes Kill Toronto Singer in Cape Breton," CBC News, October 28, 2009, https://www.cbc.ca/news/canada/nova-scotia/coyotes-kill-toronto-singer-in-cape-breton-1.779304.
197 killed by coyotes: Carly C. Sponarski, et al., "Attitudinal Differences Among Residents, Park Staff, and Visitors Toward Coyotes in Cape Breton Highlands National Park of Canada," *Society & Natural Resources* 28, no. 7 (May 2015): 720–32, https://doi.org/10.1080/08941920.2015.1014595.
197 staff or tourists were: Sponarski, et al., "Attitudinal Differences Among Residents."
198 real risk of those things: Sponarski, et al., "Attitudinal Differences Among Residents."
198 haunted the residents: Sponarski, et al., "Attitudinal Differences Among Residents."

CHAPTER 8: A FLUTTER OF SPARROWS

202 "were extremely poor": Weimin Xiong, "The 1950s Eliminate Sparrows Campaign."
202 "most easily be implemented": Xiong, "The 1950s Eliminate Sparrows Campaign."
202 a youth movement: Xiong, "The 1950s Eliminate Sparrows Campaign."
203 On a spring morning in 1958: Sha Yexin, "The Chinese Sparrows of 1958," EastWestSouthNorth, August 31, 1997, https://web.archive.org/web/20120808000323/http://www.zonaeuropa.com/20061130_1.htm.
203 between 310,000 and 800,000 sparrows killed: Yexin, "The Chinese Sparrows of 1958."
203 "that threatened them": Mikhail Klochko, *Soviet Scientist in Red China* (New York: F. Praeger, 1964).
203 "a woman's bloodcurdling screams": Klochko, *Soviet Scientist in Red China.*
204 from resting anywhere: "Great Leap," *People's Century*, PBS, Wednesday, June 16, 1999, https://www.pbs.org/wgbh/peoplescentury/episodes/greatleap/description.html.
204 "drop from exhaustion": Klochko, *Soviet Scientist in Red China.*
204 extinct in China: Michael McCarthy, "The Sparrow That Survived Mao's Purge," *Independent*, September 3, 2010, https://web.archive.org/web/20120723011028/http://www.independent.co.uk/environment/nature/nature_studies/nature-studies-by-michael-mccarthy-the-sparrow-that-survived-maos-purge-2068993.html.
204 and 1,367,440 sparrows: Frank Dikötter, *Mao's Great Famine: The History of China's Most Devastating Catastrophe, 1958–1962* (New York: Bloomsbury, 2010), 192.

204 In Judith Shapiro's: Judith Shapiro, *Mao's War Against Nature: Politics and the Environment in Revolutionary China* (Cambridge: Cambridge University Press, 2001), 87.

204 "infestations in the grain": Shapiro, *Mao's War Against Nature*, 87.

205 were already hungry: Dikötter, *Mao's Great Famine.*

205 twenty-eight million people: Hanyi Chen, et al., "Sparrow Slaughter and Grain Yield Reduction during the Great Famine of China," posted to SSRN April 23, 2021, http://dx.doi.org/10.2139/ssrn.3832057.

205 fifteen and forty-five million people: Vaclav Smil, "China's Great Famine: 40 Years Later," *British Medical Journal* 318, no. 7225 (December 1999): 1619–21, http://doi.org/ 10.1136/bmj.319.7225.1619. See also Chen, "Sparrow Slaughter and Grain Yield Reduction," and Xin Meng, et al., "The Institutional Causes of China's Great Famine, 1959–1961," *Review of Economic Studies* 82 (April 2015): 1568–611, http://doi.org/ 10.1093/restud/rdv016.

205 ate the living: Dikötter, *Mao's Great Famine*, 321, 322.

206 roots of the cane: "Canegrubs," Sugar Research Australia, https://sugarresearch.com.au/pest/canegrubs/, accessed May 19, 2022.

206 plantations in Puerto Rico: Rick Shine, *Cane Toad Wars* (Oakland: University of California Press, 2018).

207 cane toad trap: Richard Shine, et al., "A Famous Failure: Why Were Cane Toads an Ineffective Biocontrol in Australia?," *Conservation Science and Practice* 2, no. 12 (December 2020): e296, https://doi.org/10.1111/csp2.296.

208 Sydney and Canberra: Richard Shine, "The Ecological, Evolutionary, and Social Impact of Invasive Cane Toads in Australia," in *Invasive Species in a Globalized World*, ed. Keller, Cadotte, and Sandiford.

208 the 1988 documentary: *Cane Toads: An Unnatural History*, documentary, 1988, https://www.youtube.com/watch?v=6SBLf1tsoaw, accessed May 19, 2022.

209 without getting poisoned: Christa Beckmann and Richard Shine, "Toad's Tongue for Breakfast: Exploitation of a Novel Prey Type, the Invasive Cane Toad, by Scavenging Raptors in Tropical Australia," *Biological Invasions* 13 (2011): 1447–55, https://doi.org/10.1007/s10530-010-9903-8.

209 skin and poisonous shoulders: Marissa Parrott, et al., "Eat Your Heart Out: Choice and Handling of Novel Toxic Prey by Predatory Water Rats," *Australian Mammalogy* 42, no 2 (September 2019): 235–39, http://doi.org/ 10.1071/AM19016.

209 other poisonous species: Parrott, et al., "Eat Your Heart Out."

209 resistance to their poison: Ben Phillips and Richard Shine, "An Invasive Species Induces Rapid Adaptive Change in a Native Predator: Cane Toads and Black Snakes in Australia," *Proceedings of the Royal Society—Biological Sciences* 273 (March 2006): 1545–50, http://doi.org/10.1098/rspb.2006.3479.

209 dry Australian heat: Georgia Kosmala, et al., "Skin Resistance to Water Gain and Loss Has Changed in Cane Toads (Rhinella Marina) during Their

Australian Invasion," *Ecology and Evolution* 10, no 23 (December 2020): 13071–79, https://doi.org/10.1002/ece3.6895.

209 toads left behind: Shine, "The Ecological, Evolutionary, and Social Impact of Invasive Cane Toads."

210 show more cannibalism: Jayna DeVore, et al., "The Evolution of Targeted Cannibalism and Cannibal-Induced Defenses in Invasive Populations of Cane Toads," *Proceedings of the National Academy of Sciences* 118, no. 35 (August 2021): e2100765118, https://doi.org/10.1073/pnas.2100765118.

211 living-but-less-deadly toads: Reid Tingley, et al., "New Weapons in the Toad Toolkit: A Review of Methods to Control and Mitigate the Biodiversity Impacts of Invasive Cane Toads (*Rhinella Marina*)," *Quarterly Review of Biology* 92, no. 2 (June 2017): 129–49, http://doi.org/ 10.1086/692167.

212 thirty-one released lizards made it: G. Ward-Fear, J. Thomas, J. K. Webb, D. J. Pearson, and R.Shine, "Eliciting Conditioned Taste Aversion in Lizards: Live Toxic Prey Are More Effective Than Scent and Taste Cues Alone," *Integrative Zoology* 12, no. 2 (March 2017): 112–20, https://doi.org/10.1111/1749 -4877.12226.

213 biocontrol for the prickly pear: "The Prickly Pear Story," Queensland Government, Department of Agriculture and Fisheries, 2020, https://www.daf .qld.gov.au/__data/assets/pdf_file/0014/55301/prickly-pear-story.pdf.

213 devour those cacti: "The Prickly Pear Story."

213 combat grass carp: "Carp Herpes Virus First Step in Native Fish Recovery Says Alliance," Invasive Species Council, April 6, 2016, https://invasives.org.au /media-releases/carp-herpes-virus-first-step-native-fish-recovery-says-alliance/.

214 "leave not a particle": Jackson Perry, "'Conquered by the Sparrows': Avian Invasions in French North Africa, circa 1871–1920," *Environmental History* 25, no. 2 (April 2020).

215 284,000 sparrow nests: Perry, "'Conquered by the Sparrows.'"

216 "by the sparrows": Perry, "'Conquered by the Sparrows.'"

216 "horrendous sort of cruelty": Perry, "'Conquered by the Sparrows.'"

216 organize antisparrow campaigns: Perry, "'Conquered by the Sparrows.'"

217 sense of home: Matthew Holmes, "The Sparrow Question: Social and Scientific Accord in Britain, 1850–1900," *Journal of the History of Biology* 50 (August 2017): 645–71, https://doi.org/10.1007/s10739-016-9455-6.

217 waged war in newspapers: Holmes, "The Sparrow Question."

217 friend or foe: Michael Brodhead, "Elliott Coues and the Sparrow War," *New England Quarterly* 44, no. 3 (September 1971): 420–32.

217 United States denied it: Pierre Juin, "Clark Denounces Germ War Charges; Accuses Chinese Communists of Fabricating Statements Attributed to Captives," *New York Times*, February 24, 1953.

217 international investigation: "Report of the International Scientific Comission for the Investigation of the Facts Concerning Bacterial Warfare in Korea and China," International Scientific Commission, 1952.

217 were made up: Milton Leitenberg, "China's False Allegations of the Use of Biological Weapons by the United States during the Korean War," *Cold War International History Project*, Working Paper 78, March 2016.

217 there is still controversy: Diarmuid Jeffreys, "Dirty Little Secrets," Al Jazeera, March 17, 2010.

218 "an excellent option": Xiong, "The 1950s Eliminate Sparrows Campaign."

219 overinflated their numbers: Dikötter, *Mao's Great Famine*.

220 would not be punished: Dikötter, *Mao's Great Famine*, 309, 317.

220 huge floods to others: Shapiro, *Mao's War Against Nature*.

221 eat their fruit trees bare: Holmes, "The Sparrow Question."

221 poison, guns, and cats: Holmes, "The Sparrow Question."

CHAPTER 9: A HERD OF DEER

228 can leap eight feet: Leonard Perry, "Effective Deer Fences," Green Mountain Gardener, University of Vermont Department of Plant and Soil Science, https://pss.uvm.edu/ppp/articles/deerfences.html, accessed May 21, 2022.

228 crashing into a deer: "How Likely Are You to Have an Animal Collision," SimpleInsights, StateFarm, https://www.statefarm.com/simple-insights/auto-and-vehicles/how-likely-are-you-to-have-an-animal-collision, accessed May 21, 2022.

228 a white-tailed deer: Sophie Gilbert, et al., "Socioeconomic Benefits of Large Carnivore Recolonization Through Reduced Wildlife-Vehicle Collisions," *Conservation Letters* 10, no. 4 (July/August 2017): 431–39, https://doi.org/10.1111/conl.12280.

228 440 people die: Derrell Lyles, Department of Transportation, personal communication.

228 stomachs of wild animals: Michael Mengak and Mark Crosby, "Farmers' Perceptions of White-Tailed Deer Damage to Row Crops in 20 Georgia Counties During 2016," University of Georgia Warnell School of Forestry and Natural Resources, August 2017.

229 the nineteenth century: Jim Sterba, *Nature Wars: The Incredible Story of How Wildlife Comebacks Turned Backyards into Battlegrounds* (New York: Crown Publishers, 2013), 87.

229 roamed North America: Kurt VerCauteren, "The Deer Boom: Discussions on Population Growth and Range Expansion of the White-Tailed Deer," in *Bowhunting Records of North American Whitetail Deer*, 2nd ed., ed. Glenn Hisey and Kevin Hisey (Chatfield, MN: Pope and Young Club, 2003).

230 common food staples: Albert Gonzalez, "Seminole Food: Patterns of Indigenous Foodways in South Florida, 1855 to 1917," *New Florida Journal of Anthropology* 2, no. 2 (February 2021), https://doi.org/10.32473/nfja.v1i2.123723.

230 deer hair: Sterba, *Nature Wars*, 151.

230 the hunting season: J. W. Grandy, E. Stallman, and D. Macdonald, "The Science and Sociology of Hunting: Shifting Practices and Perceptions in

the United States and Great Britain," in *The State of the Animals II*, ed. D. J. Salem and A. N. Rowan (Washington, DC: Humane Society Press, 2003), 107–30).

230 below fifteen million: VerCauteren, "The Deer Boom."

230 in the United States: VerCauteren, "The Deer Boom."

230 to become "wild": Sterba, *Nature Wars*, 87

230 in the Adirondacks: Sterba, *Nature Wars*, 98.

231 functionally illegal: Sterba, *Nature Wars*, 95.

231 The state agencies do: Sterba, *Nature Wars*, 103.

231 live in the woods already: Sterba, *Nature Wars*.

231 "two thirds of the U.S. population": Sterba, *Nature Wars*, 2.

232 habitat Americans made: VerCauteren, "The Deer Boom."

232 forty-five deer: Quality Deer Management Association, QDMA Whitetail Report, Bogart, GA, 2009.

233 from the ground up: Keith Geluso, Carter G. Kruse, and Mary J. Harner, "Wetland Edge Trampled by American Bison (Bos Bison) Used as Basking Site for Painted Turtles (*Chrysemys picta*)," *Transactions of the Nebraska Academy of Sciences and Affiliated Societies* (2020), https://digitalcommons.unl.edu/cgi /viewcontent.cgi?article=1543&context=tnas.

233 with large grazers: Geluso, Kruse, and Harner, "Wetland Edge Trampled by American Bison."

233 to sun on lakeshores: Geluso, Kruse, and Harner, "Wetland Edge Trampled by American Bison."

233 colonists in 1788: "Australian Brumby Horse," Breeds of Livestock, Department of Animal Sciences, Oklahoma State University, http://afs.okstate.edu/breeds /horses/australianbrumby, accessed May 21, 2022.

233 feral by 1804: "Feral Horse (*Equus caballus*) and Feral Donkey (*Equus asinus*)," Fact Sheet, Australian Government, Department of Agriculture, Water, and the Environment, 2011.

233 feral horses roam Australia: "Feral Horses," Invasive Species Council, https:// invasives.org.au/our-work/feral-animals/feral-horses/, accessed May 21, 2022.

234 at their heels: Banjo Paterson, "The Man from Snowy River."

234 with gunshot wounds: Biance Nogrady, "In Australia's Snowy Mountains, a Battle Over Brumbies," Undark, July 25, 2018, https://undark.org/2018/07/25 /battle-over-brumbies/.

234 has banned the practice: Nogrady, "In Australia's Snowy Mountains, a Battle Over Brumbies."

235 lizards, and freshwater crayfish: Rosie King, "Kosciuszko National Park Brumbies Causing 'Abhorrent' Damage, Says Indigenous River Guide," ABC News (Australia), July 6, 2020, https://www.abc.net.au/news/2020-07-07 /kosciuszko-feral-horses-controversy/12405310.

236 continued to increase: Don Driscoll, et al., "Feral Horses Will Rule One Third of the Fragile Kosciuszko National Park under a Proposed NSW Government Plan," The Conversation, October 7, 2021, https://theconversation.com/feral

-horses-will-rule-one-third-of-the-fragile-kosciuszko-national-park-under-a
-proposed-nsw-government-plan-169248.

238 horses and frogs alike: Don A. Driscoll and Maggie J. Watson, "Science
Denialism and Compassionate Conservation: Response to Wallach et al. 2018,"
Conservation Biology 33, no. 4 (August 2019): 777–80, https://doi.org/10.1111
/cobi.13273.

239 the vibrant hue: Margaret Davis, "Australia Culling 10,000 Feral Horses to
Control Their Rapidly Growing Population Threatening Endangered Species,
Habitats," *Science Times*, November 1, 2021, https://www.sciencetimes.com
/articles/34272/20211101/australia-culling-10-000-feral-horses-control-rapidly
-growing-population.htm.

239 They can even see orange: Kurt VerCauteren and Michael Pipas, "A Review
of Color Vision in White-Tailed Deer," *Wildlife Society Bulletin* 31, no. 3
(Autumn 2006): 684–91.

242 lower deer densities: Donald Waller and Nicholas Reo, "First Stewards:
Ecological Outcomes of Forest and Wildlife Stewardship by Indigenous Peoples
of Wisconsin, USA," *Ecology and Society* 23, no. 1 (2018): 45, https://doi
.org/10.5751/ES-09865-230145.

242 deer are native: Sterba, *Nature Wars.*

243 epizootic hemorrhagic disease: Sonja A. Christensen, et al., "Spatial
Variation of White-Tailed Deer (*Odocoileus Virginianus*) Population Impacts
and Recovery from Epizootic Hemorrhagic Disease," *Journal of Wildlife Diseases* 57, no. 1 (January 2021): 82–93, https://doi.org/10.7589/JWD-D-20
-00030.

243 chronic wasting disease: Michael Samuel and Daniel Storm, "Chronic Wasting
Disease in White-Tailed Deer: Infection, Mortality, and Implications for
Heterogeneous Transmission," *Ecology* 97, no. 11 (November 2016): 3195–205,
https://doi.org/10.1002/ecy.1538.

243 even COVID-19: Mitchell Palmer, et al., "Susceptibility of White-Tailed Deer
(*Odocoileus virginianus*) to SARS-CoV-2," *Journal of Virology* 95, no. 11 (2021),
https://doi.org/10.1128/JVI.00083-21.

243 sharpshooters in 2004: Thomas Almendinger, et al., "Restoring Forests in
Central New Jersey Through Effective Deer Management," *Ecological
Restoration* 38, no. 4 (December 2020): 246–56, http://doi.org/10.3368/
er.38.4.246.

243 more abundant outside: Almendinger, "Restoring Forests in Central New
Jersey."

244 in the United States alone: "Lyme Disease: Data and Surveillance," Centers for
Disease Control and Prevention, https://www.cdc.gov/lyme/datasurveillance
/index.html?CDC_AA_refVal=https%3A%2F%2Fwww.cdc
.gov%2Flyme%2Fstats%2Findex.html, accessed May 21, 2022.

244 Lyme, Connecticut: "A Brief History of Lyme Disease in Connecticut"
Connecticut State Department of Health, updated July 1, 2019. https://portal

.ct.gov/DPH/Epidemiology-and-Emerging-Infections/A-Brief-History-of-Lyme
-Disease-in-Connecticut.

245 "Lyme arthritis": "A Brief History of Lyme Disease in Connecticut."

245 deer—were limited: A. M. Kilpatrick, et al., "Lyme Disease Ecology in a
Changing World: Consensus, Uncertainty and Critical Gaps for Improving
Control," *Philosophical Transactions of the Royal Society B* 372, no. 1722 (April
2017), http://dx.doi.org/10.1098/rstb.2016.0117.

245 3.6 per doe: Uriel Kitron, et al., "Spatial Analysis of the Distribution of *Ixodes
dammini* (Acari: Ixodidae) on White-Tailed Deer in Ogle County, Illinois,"
Journal of Medical Entomology 29, no. 2 (March 1992): 259–66, https://doi
.org/10.1093/jmedent/29.2.259.

245 12.7 ticks per deer: M. L. Baer-Lehman, et al., "Evidence for Competition
between *Ixodes scapularis* and *Dermacentor albipictus* Feeding Concurrently on
White-Tailed Deer," *Experimental and Applied Acarology* 58 (May 2012): 301–
14, https://doi.org/10.1007/s10493-012-9574-5.

245 555 larval ticks per deer: Ching-I Huang, et al., "High Burdens of *Ixodes
scapularis* Larval Ticks on White-Tailed Deer May Limit Lyme Disease Risk in
a Low Biodiversity Setting," *Ticks and Tick-borne Diseases* 10, no. 2 (February
2019): 258–68, https://doi.org/10.1016/j.ttbdis.2018.10.013.

245 doing controlled burns: Kilpatrick, et al., "Lyme Disease Ecology in a
Changing World."

245 fears of side effects and low demand: L. E. Nigrovic and K. M. Thompson,
"The Lyme Vaccine: A Cautionary Tale," *Epidemiology & Infection* 135,
no. 1 (January 2007): 1–8, http://doi.org/ 10.1017
/S0950268806007096.

245 hope for another: Cassandra Willyard, "Lyme Vaccines Show New Promise,
and Face Old Challenges," Undark, October 2, 2019, https://undark
.org/2019/10/02/new-landscape-lyme-vaccines/.

246 in the Reservation: John Patrick Connors and Anne Short Gianotti,
"Becoming Killable: White-Tailed Deer Management and the Production of
Overabundance in the Blue Hills," *Urban Geography* (May 2021), https://doi
.org/10.1080/02723638.2021.1902685.

246 a controlled hunt: Connors and Gianotti, "Becoming Killable."

247 some several times: Connors and Gianotti, "Becoming Killable."

247 Friends of the Blue Hills: Connors and Gianotti, "Becoming Killable."

248 on Staten Island: Brooke Jarvis, "Deer Wars and Death Threats," *New Yorker*,
November 8, 2021, https://www.newyorker.com/magazine/2021/11/15/deer
-wars-and-death-threats.

248 National Institutes of Health: "Fertility Control," White Buffalo, Inc., https://
www.whitebuffaloinc.org/fertility-control, accessed May 21, 2022.

248 twenty-three deer: Blue Hills State Reservation, White-Tailed Deer
Management Program, 2021 Deer Management Plan, October 8, 2021.

249 deer became killable: Connors and Gianotti, "Becoming Killable."

CHAPTER 10: A SLOTH OF BEARS

256 and Kentucky: Sean M. Murphy, et al., "Early Genetic Outcomes of American Black Bear Reintroductions in the Central Appalachians, USA," *Ursus* 29, no. 2 (May 2019): 119–33, https://doi.org/10.2192/URSU-D-18-00011.1.

258 got a bear visit: Stewart W. Breck, Nathan Lance, and Victoria Seher, "Selective Foraging for Anthropogenic Resources by Black Bears: Minivans in Yosemite National Park," *Journal of Mammalogy* 90, no. 5 (October 2009): 1041–44, https://doi.org/10.1644/08-MAMM-A-056.1.

258 forty years of yearling bears: Nicholas P. Gould, et al., "Growth and Reproduction by Young Urban and Rural Black Bears," *Journal of Mammalogy* 102, no. 4 (August 2021): 1165–73, https://doi.org/10.1093/jmammal/gyab066.

259 bred at that age: Gould, et al., "Growth and Reproduction."

259 new set of skills: B. Brookshire, "Changing People's Behavior Can Make Bear Life Better," *Science News for Students*, April 8, 2021, https://www.sciencenewsforstudents.org/article/changing-people-behavior-can-make-bear-life-better.

259 "landscape of fear": Kathy Zeller, et al., "Black Bears Alter Movements in Response to Anthropogenic Features with Time of Day and Season," *Movement Ecology* 7, no. 19 (July 2019), https://doi.org/10.1186/s40462-019-0166-4.

259 11:00 p.m. to 3:00 a.m.: Lee Anne Ayers, et al., "Black Bear Activity Patterns and Human Induced Modifications in Sequoia National Park," *Bears: Their Biology and Management* 6 (1986): 151–54, https://doi.org/10.2307/3872819.

259 spring and fall: Zeller, "Black Bears Alter Movements."

260 heart rates increased: Mark A. Ditmer, et al., "American Black Bears Perceive the Risks of Crossing Roads," *Behavioral Ecology* 29, no. 3 (May/June 2018): 667–75, https://doi.org/10.1093/beheco/ary020.

260 Tahoe, and Durango: H. E. Johnson, et al., "Shifting Perceptions of Risk and Reward: Dynamic Selection for Human Development by Black Bears in the Western United States," *Biological Conservation* 187 (July 2015): 164–72, https://doi.org/10.1016/j.biocon.2015.04.014.

261 2.3 days shorter: Heather Johnson, et al., "Human Development and Climate Affect Hibernation in a Large Carnivore with Implications for Human–Carnivore Conflicts," *Journal of Applied Ecology* 55, no. 2 (2017): 663–72, https://doi.org/10.1111/1365-2664.13021.

261 human and natural food: Johnson, et al., "Human Development and Climate Affect Hibernation."

261 by 60 percent: Rebecca Kirby, et al., "The Diet of Black Bears Tracks the Human Footprint across a Rapidly Developing Landscape," *Biological Conservation* 200 (August 2016): 51–59, https://doi.org/10.1016/j.biocon.2016.05.012.

262 slows that trimming: Rebecca Kirby, et al., "The Cascading Effects of Human Food on Hibernation and Cellular Aging in Free-Ranging Black Bears,"

Scientific Reports 9, no. 2197 (February 2019), https://doi.org/10.1038/s41598 -019-38937-5.

262 aging more rapidly: Kirby, et al., "The Cascading Effects of Human Food."

262 human-caused death: Jared S. Laufenberg, et al., "Compounding Effects of Human Development and a Natural Food Shortage on a Black Bear Population along a Human Development-Wildland Interface," *Biological Conservation* 224 (August 2018): 188–98, https://doi.org/10.1016/j.biocon.2018.05.004.

263 for bear roadkills: Sharon Baruch-Mordo, et al., "Spatiotemporal Distribution of Black Bear–Human Conflicts in Colorado, USA," *Journal of Wildlife Management* 72, no. 8 (November 2008): 1853–62, https://doi.org/10.2193 /2007-442.

263 bear-resistant trash cans: Heather Johnson, et al., "Assessing Ecological and Social Outcomes of a Bear-Proofing Experiment," *Journal of Wildlife Management* 82, no. 6 (August 2018): 1102–14, https://doi.org/10.1002 /jwmg.21472.

264 half actually did: Stacy A. Lischka, et al., "A Conceptual Model for the Integration of Social and Ecological Information to Understand Human-Wildlife Interactions," *Biological Conservation* 225 (September 2018): 80–87, https://doi.org/10.1016/j.biocon.2018.06.020.

265 used them correctly: Stacy A. Lischka, et al., "Psychological Drivers of Risk-Reducing Behaviors to Limit Human–Wildlife Conflict," *Conservation Biology* 34, no. 6 (December 2020): 1383–92, https://doi.org/10.1111/cobi.13626.

265 39 percent more: Johnson, et al., "Assessing Ecological and Social Outcomes."

268 too close to people: Alyze Kotyk, "Humans' Behaviour May Change if They Realized How Many Black Bears Are Killed Every Year in B.C.: Advocate," CTV News, Vancouver, Canada, January 20, 2022, https://bc.ctvnews.ca /humans-behaviour-may-change-if-they-realized-how-many-black-bears-are -killed-every-year-in-b-c-advocate-1.5747815.

268 Virginia City in Montana: Michael Cast, "Virginia City: The Model Bear-Smart Community," *Missoulian*, January 9, 2022, https://missoulian.com/news /local/virginia-city-the-model-bear-smart-community/article_06cd732b-7dfe -55db-bf77-419f0fbaf711.html.

269 damaging their property: Stacy A. Lischka, et al., "Understanding and Managing Human Tolerance for a Large Carnivore in a Residential System," *Biological Conservation* 238 (October 2019), doi: 10.1016/j.biocon.2019 .07.034.

CHAPTER 11: A PEST BY ANY OTHER NAME

274 "pest control operator" in 1936: Sullivan, *Rats*.

275 "matter out of place": Mary Douglas, *Purity and Danger: An Analysis of Concept of Pollution and Taboo* (New York: Routledge & Kegan Paul, 1966).

276 Genesis 1:28: Richard Wright, "Responsibility for the Ecological Crisis," *BioScience* 20, no. 15 (August 1970): 851–53, https://doi.org/10.2307/1295493.

276 upon the earth: Genesis 1:28, https://www.kingjamesbibleonline.org/Genesis -1-28/.

278 "mutuality of life": Marshall, *On Behalf of the Wolf and the First Peoples*.

279 "key to survival": Marshall, *On Behalf of the Wolf and the First Peoples*, 8.

280 "we do the same": Sullivan, *Rats*, 2.

287 responded with panic: B. Brookshire, "As Wild Turkeys Grow in Number, so Do Risky Encounters with Humans," *Washington Post*, April 22, 2022, https:// www.washingtonpost.com/health/2022/04/22/wild-turkey-anacostia-attacks/.

287 as early as 300 BCE: Erin Kennedy Thornton, et al., "Earliest Mexican Turkeys (*Meleagris gallopavo*) in the Maya Region: Implications for Pre-Hispanic Animal Trade and the Timing of Turkey Domestication," *PLOS ONE* 7, no. 8 (August 2012): e42630, https://doi.org/10.1371/journal.pone.0042630.

287 the Civil War: Sterba, *Nature Wars*, 151.

288 flocks in the United States took off: Sterba, *Nature Wars*, 151.

288 four million birds: James E. Miller, et al., "Turkey Damage Survey: A Wildlife Success Story Becoming Another Wildlife Damage Problem," *Wildlife Damage Management Conference Proceedings*, October 2000.

288 ruined by scratching turkey claws: James E. Miller, "Wild Turkeys," Wildlife Damage Management Technical Series, USDA, APHIS, WS, January 2018.

288 cannot, in fact, solve everything: Joshua Bote, "Destructive Turkeys Are Creating a Nightmare at Bay Area NASA Lab," SFGate, February 8, 2022, https://www.sfgate.com/bayarea/article/Turkeys-create-nightmare-NASA-Ames -lab-16841998.php.

FURTHER READING

This is not a full bibliography of everything I've ever read (trust me, no one wants that). Most specific citations that are important I've left to the endnotes. Instead, this is a list of books, articles, and more that helped to form my thinking about *Pests*, and that will give you more amazing information about the animals that love to live with us.

HUMAN-WILDLIFE INTERACTIONS

Barilla, James. *My Backyard Jungle: The Adventures of an Urban Wildlife Lover Who Turned His Yard into Habitat and Learned to Live with It*. New Haven: Yale University Press, 2013.

Biehler, Dawn. *Pests in the City: Flies, Bedbugs, Cockroaches & Rats*. Seattle: University of Washington Press, 2013.

Dikötter, Frank. *Mao's Great Famine: The History of China's Most Devastating Catastrophe, 1958–1962*. New York: Bloomsbury, 2010.

Donovan, Tristan. *Feral Cities: Adventures with Animals in the Urban Jungle*. Chicago: Chicago Review Press, 2015.

Hanes, Stephanie. *White Man's Game: Saving Animals, Rebuilding Eden, and Other Myths of Conservation in Africa*. New York: Metropolitan Books, 2017.

Herzog, Hal. *Some We Love, Some We Hate, Some We Eat: Why It's So Hard to Think Straight About Animals*. New York: HarperCollins, 2010.

Johnson, Nathanael. *Unseen City: The Majesty of Pigeons, the Discreet Charm of Snails & Other Wonders of the Urban Wilderness*. New York: Rodale Wellness, 2016.

Kimmerer, Robin Wall. *Braiding Sweetgrass: Indigenous Wisdom, Scientific Knowledge, and the Teachings of Plants*. Minneapolis, MN: Milkweed Editions, 2013.

King, Barbara J. *Animals' Best Friends: Putting Compassion to Work for Animals in Captivity and in the Wild*. Chicago: Chicago University Press, 2021.

Marris, Emma. *Rambunctious Garden: Saving Nature in a Post-Wild World*. New York: Bloomsbury, 2011.

————. *Wild Souls: Freedom and Flourishing in the Non-Human World*. New York: Bloomsbury, 2021.

Marshall, Joseph, III. *On Behalf of the Wolf and the First Peoples*. Santa Fe: Museum of New Mexico Press, 1995.

Pavlik, Steve. *Navajo and the Animal People: Native American Traditional Ecological Knowledge and Ethnozoology*. Golden, CO: Fulcrum Publishing, 2014.

Roach, Mary. *Fuzz: When Nature Breaks the Law*. New York: W. W. Norton & Company, 2021.

Shapiro, Judith. *Mao's War Against Nature: Politics and the Environment in Revolutionary China*. Cambridge: Cambridge University Press, 2001.

Shivik, John A. *The Predator Paradox: Ending the War with Wolves, Bears, Cougars, and Coyotes*. Boston: Beacon Press, 2014.

Sterba, Jim. *Nature Wars: The Incredible Story of How Wildlife Comebacks Turned Backyards into Battlegrounds*. New York: Crown, 2013.

SPECIES-SPECIFIC BOOKS AND ARTICLES

Blechman, Andrew D. *Pigeons: The Fascinating Saga of the World's Most Revered and Reviled Bird*. New York: Grove Press, 2006.

Bradshaw, John. *Cat Sense: How the New Feline Science Can Make You a Better Friend to Your Pet*. New York: Basic Books, 2013.

Brookshire, Bethany. "As Wild Turkeys Grow in Number, So Do Risky Encounters with Humans." *Washington Post*, April 22, 2022. https://www.washingtonpost .com/health/2022/04/22/wild-turkey-anacostia-attacks/.

————. "Changing People's Behavior Can Make Bear Life Better." *Science News for Students*, April 8, 2021. https://www.sciencenewsforstudents.org/article /changing-people-behavior-can-make-bear-life-better.

————. "How the House Mouse Tamed Itself." *Science News*, April 19, 2017. https:// www.sciencenews.org/blog/scicurious/how-house-mouse-tamed-itself.

Burt, Jonathan. *Rat*. London: Reaktion Books, 2005.

"Cane Toads: An Unnatural History." YouTube. Posted May 11, 2015. https://youtu .be/6SBLfl tsoaw.

Cecco, Leyland. "Raccoons v Toronto: How 'Trash Pandas' Conquered the City." *Guardian*, October 5, 2018. https://www.theguardian.com/world/2018/oct/05 /canada-toronto-raccoons.

Coleman, Jon T. *Vicious: Wolves and Men in America*. New Haven, CT: Yale University Press, 2004.

Flores, Dan. *Coyote America: A Natural and Supernatural History*. New York: Basic Books, 2016.

Hongoltz-Hetling, Matthew. *A Libertarian Walks into a Bear: The Utopian Plot to Liberate an American Town (And Some Bears)*. New York: PublicAffairs, 2020.

Humphries, Courtney. *Superdove: How the Pigeon Took Manhattan . . . and the World*. New York: HarperCollins, 2008.

Jarvis, Brooke. "Deer Wars and Death Threats." *New Yorker*, November 8, 2021. https://www.newyorker.com/magazine/2021/11/15/deer-wars-and-death-threats.

Jerolmack, Colin. *The Global Pigeon*. Chicago: University of Chicago Press, 2013.

Justice, Daniel Heath. *Raccoon*. London: Reaktion Books, 2021.

Marra, Peter P., and Chris Santella. *Cat Wars: The Devastating Consequences of a Cuddly Killer*. Princeton, NJ: Princeton University Press, 2016.

National Geographic Society. "How European Rabbits Took Over Australia." National Geographic.org. Posted on January 27, 2020. https://www .nationalgeographic.org/article/how-european-rabbits-took-over-australia/.

Nograby, Bianca. "In Australia's Snowy Mountains, a Battle over Brumbies." Undark, July 25, 2018. https://undark.org/2018/07/25/battle-over-brumbies/.

Shine, Rick. *Cane Toad Wars*. Oakland: University of California Press, 2018.

Stolzenburg, William. *Rat Island: Predators in Paradise and the World's Greatest Wildlife Rescue*. New York: Bloomsbury, 2011.

Sullivan, Robert. *Rats: Observations on the History & Habitat of the City's Most Unwanted Inhabitants*. New York: Bloomsbury, 2004.

Vann, Michael, and Liz Clarke. *The Great Hanoi Rat Hunt: Empire, Disease, and Modernity in French Colonial Vietnam*. New York: Oxford University Press, 2019.

HUMAN-WILDLIFE INTERACTIONS: SCIENTIFIC ARTICLES AND BOOKS

Douglas, Mary. *Purity and Danger: An Analysis of Concepts of Pollution and Taboo*. New York: Routledge & Kegan Paul, 1966.

Frank, Beatrice, Jenny A. Glikman, and Silvio Marchini, eds. *Human Wildlife Interactions: Turning Conflict into Coexistence*. Cambridge: Cambridge University Press, 2019.

Mikhail, Alan. *The Animal in Ottoman Egypt*. Oxford: Oxford University Press, 2014.

Nyhus, Phillip J. "Human-Wildlife Conflict and Coexistence." *Annual Review of Environment and Resources* 41 (2016): 143–71. https://doi.org/10.1146/annurev -environ-110615-085634.

Ritvo, Harriet. *The Animal Estate: The English and Other Creatures in the Victorian Age*. Cambridge, MA: Harvard University Press, 1987.

Stone, Ken. *Reading the Hebrew Bible with Animal Studies*. Stanford, CA: Stanford University Press, 2018.

SPECIES-SPECIFIC SCIENTIFIC ARTICLES AND BOOKS

Haami, Bradford. *Cultural Knowledge and Traditions Relating to the Kiore Rat in Aotearoa. Part 1: A Maori Perspective*. Science and Mathematics Education Papers. Hamilton, NZ: University of Waikato, 1993.

Holmes, Matthew. "The Perfect Pest: Natural History and the Red Squirrel in Nineteenth-Century Scotland." *Archives of Natural History* 2, no. 1 (2015): 113–25. https://doi.org/10.3366/anh.2015.0284.

LoBue, Vanessa, and Karen E. Adolph. "Fear in Infancy: Lessons from Snakes, Spiders, Heights, and Strangers." *Developmental Psychology* 55, no. 9 (2019): 1889–907. https://doi.org/1010.1037/dev0000675.

Nelson, Nicole C. *Model Behavior: Animal Experiments, Complexity, and the Genetics of Psychiatric Disorders*. Chicago: University of Chicago Press, 2018.

Rader, Karen. *Making Mice: Standardizing Animals for American Biomedical Research, 1900–1955*. Princeton, NJ: Princeton University Press, 2004.

Roberts, Mere. *Scientific Knowledge and Cultural Traditions. Part 2: A Pakeha View of the Kiore Rat in New Zealand*. Science and Mathematics Education Papers. Hamilton, NZ: University of Waikato, 1993.

Serpall, James A. "Domestication and History of the Cat." In *The Domestic Cat: The Biology of its Behaviour*, edited by Dennis C. Turner and Patrick Bateson, 83–100. Cambridge: Cambridge University Press, 2013.

INDEX

Page numbers with illustrations are in *italics*.